LEE KREUTZER

Bison Kills and Bone Counts

PREHISTORIC ARCHEOLOGY AND ECOLOGY
A Series Edited by Karl W. Butzer and Leslie G. Freeman

Bison Kills and Bone Counts

Decision Making by Ancient Hunters

John D. Speth

University of Chicago Press
Chicago and London

JOHN D. SPETH is the author of numerous archeological research reports. He is currently associate professor of anthropology and associate curator of North American archeology at the University of Michigan.

The University of Chicago Press, Chicago 60637

The University of Chicago Press, Ltd., London

Printed in the United States of America

90 89 88 87 86 85 84 83 5 4 3 2 1

Library of Congress Cataloging in Publication Data

Speth, John D.
  Bison kills and bone counts.

  (Prehistoric archeology and ecology)
  Bibliography: p.
  Includes index.
  1. Garnsey Site (N.M.) 2. Indians of North
America--Southwest, New--Food. 3. Animal remains
(Archaeology)--New Mexico. I. Title. II. Series.
E78.N65S65 1983   978.9'43   82-21976
ISBN 0-226-76948-8
ISBN 0-226-76949-5 (pbk.)

To my parents,
Alfred and Lottie,
who introduced me to the
beauty and complexity of nature

Contents

Series Editors' Foreword

John Speth takes a methodological approach to interpreting the Garnsey bison kill, a well-preserved and documented kill site dating from only a century or so before the introduction of the horse to the Great Plains.

He shows that the bison parts we should expect to find in large numbers at a kill site are those with least nutritional value. Based on these premises, he applies "nutritional analysis" to a real bison kill, a method that, as far as we know, has been applied only once before to archeological materials. That was in Lewis Binford's seminal <u>Bones</u>: <u>Ancient</u> <u>Men</u> <u>and</u> <u>Modern</u> <u>Myths</u>, devoted primarily to the caribou. Although Speth in some respects follows Binford's lead, his development of the method improves upon Binford's, and the greater clarity achieved should do much to establish its potential.

Speth goes further than Binford by emphasizing that the nutritional value of a body part does not remain constant throughout the year, but varies with the season. He also

notes that in a species like bison such seasonal variation
will differ between the sexes. The implication is that,
depending on the season in which a kill is made, the bones
of less valued body parts from one sex will tend to be left
at the site out of proportion to the representation of that
sex in the kill sample; that is, the sex ratio as calculated
from those bones will come to differ from the sex ratio in
the live (or just killed) animals. The sex ratio may also
be different for each body part in the sample left behind at
the kill site: only those bones that have virtually no nutritional value in either sex will reflect the original sex
ratio of the animals taken.

An important implication for zooarcheologists, whether
they are working with bison or with other ungulate species,
is that they must calculate the sex ratio from a variety of
body elements, not simply from the one that is most con-
venient. Speth's application of the "nutritional utility"
method to the Garnsey bison bones, combined with his detailed analysis of their sex and age composition, makes this
volume unique. We believe that not only fauna specialists
but all archeologists concerned with procurement methods and
butchering/processing procedures will find <u>Bison Kills and
Bone Counts</u> of fundamental value.

Karl W. Butzer

Leslie G. Freeman

Preface

The object of this study is to develop an understanding of the importance of animal physiology in the procurement and processing decisions of prehistoric bison hunters. Archeologists have long recognized the significance of condition in terms of which animals were likely to be hunted. Thus most would agree that seasonal differences in the condition of males and females favored preferential procurement of bulls in the spring and cows in the fall and winter. However, archeologists have largely ignored animal condition as a factor affecting subsequent processing decisions, assuming instead that all animals, once killed, would have been handled in more or less the same manner. This study attempts to show that processing decisions by prehistoric hunters were necessarily highly selective, and that in certain circumstances the physiological condition or fat content of an anatomical part may have been as important as its bulk in determining whether it would be removed from a kill or abandoned.

The research that forms the heart of this study began
with the investigation of a seemingly anomalous late prehistoric bison kill (Garnsey Site) in the Southern Plains region of New Mexico. What made the kill unusual was that it
consisted of a series of spring-season events in which the
preferred targets were bulls, not cows. This contrasts
rather strikingly with the well-known Northern Plains pattern in which cows were the principal focus of procurement
activities and hunting took place mostly in the fall and
winter. Not only were bulls the major target at the Garnsey
Site, but many of the cows that were killed were systematically discriminated against during butchering and processing. As a result, most of the bones left behind at the site
were from cows, not bulls.

Closer examination of the abandoned skeletal parts
showed that discrimination at Garnsey was directed specifically against animals that were in poorest overall condition
in the spring; that is, pregnant or lactating cows. Moreover, discrimination was strongest against anatomical parts,
such as marrow bones, which were most vulnerable to fat depletion. These and other observations indicated that the
condition of the animals, particularly their level of body
fat, was indeed an extremely important consideration in the
processing decisions at Garnsey; but it remained to be determined why.

To better understand the importance of fat in the diets of hunting populations, I turned to contemporary ethnographic and ethnohistoric accounts. From these it immediately became apparent that fatty foods were highly valued by virtually all hunters and gatherers, regardless of latitude. More interesting, however, were numerous cases, similar to what had been observed at Garnsey, in which hunters avoided or abandoned animals they considered too lean for use, even when the hunters themselves were short of food. This suggested that fat was valued not merely for its desirable taste, but also for more important underlying nutritional reasons. The next step obviously was to explore the nutritional role of fat in hunter-gatherer diets.

Fat is, above all, a highly concentrated source of energy. In addition, at high overall calorie intakes, both fat and carbohydrate exert a significant and roughly comparable protein-sparing effect during protein metabolism. Put another way, the body makes more effective use of dietary protein when energy is provided by nonprotein sources, regardless of whether the calories derive from fat or carbohydrate. However, at low total calorie intakes the protein-sparing capacities of fat and carbohydrate diverge significantly. In other words, as total energy intake declines, the body loses protein more rapidly if energy is supplied by fat rather than by carbohydrate.

If these observations stand up to further scrutiny, they have important implications for understanding not only the highly selective behavior witnessed at Garnsey, but also several other seemingly enigmatic aspects of hunter-gatherer subsistence behavior. Hunters and gatherers regularly face periods of restricted energy intake in late winter and spring. At such times they often subsist on stored carbohydrate foods and supplement their diet with hunted foods. It is precisely at such times that the level of fat in the diet becomes critical. To make effective use of the protein provided by hunting, hunters must maintain their calorie intake from fat as high as possible or else find alternative sources of carbohydrate. Briefly, some of the more obvious options open to hunters and gatherers to cope with the noninterchangeability of fat and carbohydrate at low total energy intakes include: (1) being highly selective in the animals they kill and the parts they consume (as at Garnsey); (2) switching to species that maintain high body-fat levels throughout the winter and spring (e.g., beaver or geese); (3) emphasizing plant gathering rather than hunting in the fall, in order to build up large carbohydrate reserves; (4) undertaking limited cultivation; or (5) trading for carbohydrates with horticulturalists.

Clearly, a great deal of additional research needs to be done to reliably model seasonal changes in animal condition, and especially to arrive at a satisfactory understanding of the effect of these changes on the food-getting strategies of prehistoric hunters and gatherers. I hope the Garnsey study will serve as a first step toward these goals.

Acknowledgments

The two seasons of fieldwork at the Garnsey Bison Kill Site were conducted under Federal Antiquities Act permit 77-NM-037, issued to the University of Michigan by the United States Department of the Interior and administered by the Bureau of Land Management in the State of New Mexico (Roswell District). Additional work of a limited nature was conducted at the site during the summer of 1981, under temporary Federal Antiquities Act permit 81-EM-036.

I am extremely grateful to Elmer and Jane Garnsey, on whose ranch the bison kill is situated, for their generosity and kindness throughout the project.

Much of the funding for the research was provided by the National Science Foundation (grant BNS-7806875). Additional funds were provided by the Horace H. Rackham School of Graduate Studies and by the Museum of Anthropology, both of the University of Michigan.

Figures 1-20 and 27-28 of this study are reproduced from Speth and Parry (1980) with the permission of the

Museum of Anthropology, University of Michigan. Figure 26 is reproduced with Michael Wilson's permission from his contribution in the same report. Figures 21-25, 33, and 36-60 were drawn by Lisa Klofkorn.

Earlier versions of this manuscript were read and critically commented upon by Karl W. Butzer, Richard I. Ford, Richard G. Klein, Katherine M. Moore, Susan L. Scott, Katherine A. Spielmann, and Wirt H. Wills, to whom I am most grateful. Obviously, whatever shortcomings remain in the study, and there undoubtedly are many, are entirely my own.

Abbreviations

The following abbreviations are frequently used in the text and figures.

Anatomical Parts

Ast	Astragalus	Pel	Pelvis
At	Atlas	PF	Proximal femur
Ax	Axis	PH	Proximal humerus
Calc	Calcaneus	Ph 1	First phalanx
Car	Carpal	Ph 2	Second phalanx
Cer	Cervical	Ph 3	Third phalanx
DF	Distal femur	PMc	Proximal metacarpal
DH	Distal humerus	PMt	Proximal metatarsal
DMc	Distal metacarpal	PR	Proximal radius
DMt	Distal metatarsal	PT	Proximal tibia
DR	Distal radius	R	Radius
DT	Distal tibia	Scap	Scapula
F	Femur	Sk	Skull
H	Humerus	Ster	Sternum
Lum	Lumbar	T	Tibia
Man	Mandible	Tar	Tarsal
Mc	Metacarpal	Th	Thoracic
Mt	Metatarsal	U	Ulna

Others

GUI	General utility index
MGUI	Modified general utility index
MNI	Minimum number of individuals
%MNI	Percentage (of maximum value of) minimum number of individuals

1. Introduction

Plains hunter-gatherers exercised considerable selectivity in hunting bison of a particular sex (cf. Ewers 1958:76; Wissler 1910:41; Point 1967:120, 166; Grinnell 1972:269; see also Appendix P in Roe 1972:860-61). An obvious reason for such selectivity, and the one most widely discussed in the literature, stems from behavioral characteristics that made bison of one sex easier to drive or manipulate at particular times of year than animals of the other sex (e.g., Wheat 1972; Frison 1974, 1978; Arthur 1975). Thus, for example, cows with calves were difficult to handle in the spring because the calves' behavior was unpredictable, and a mother usually followed her calf if it bolted from the herd. Smaller, more stable bull groups were common targets in the spring. During the late-summer rut bulls became much more aggressive and unstable and thus more difficult to drive. Their disruptive behavior at this time of year also made females harder to handle. Cow groups became the principal focus of hunting in the fall and winter,

once the bulls had withdrawn from the herds after the rut
and the calves had matured enough that their behavior was no
longer erratic.

Plains procurement strategies were also clearly keyed
to differences in the nutritional condition of cows and
bulls at different seasons. For example, cows were commonly
avoided in the spring when their condition was poorest but·
became prime targets in the fall and early winter as they
built up reserves of fat to carry them through the winter.
Males, on the other hand, were preferred prey in the late
spring and early summer, when they were in better condition
than the cows, but were avoided during and after the rut
when they often were in poorer shape (Ewers 1958:76; Wissler
1910:41; Wilson 1924:306; Point 1967:120, 166; Grinnell
1972:269; Earl of Southesk 1969:80; Coues 1897:577; Roe
1972:860-61; Lott 1979:429).

Finally, Plains hunter-gatherers chose bison of a particular sex to provide hides for clothing and shelter, and
secondarily to obtain specific by-products of the kill for a
wide range of tools, ritual paraphernalia, and so forth
(Ewers 1955:150, 1958:76). For example, they used cow skins
for robes; lodge covers, doors, and linings; bedding; clothing; and a variety of leather containers. They sometimes
killed cows in the spring to obtain fetuses, which, in addition to being considered a delicacy by many Plains groups,
were valued for making soft bags and children's clothing.

Thicker, tougher bull hides were used to make sinew for bow-strings and bow backings; rawhide strips and thongs; shield covers; parfleches; bedding; glue; moccasin soles; and so forth. Some groups took bull tongues for ritual feasting associated with the Sun Dance (Ewers 1955:150, 1958:74-76, 90; Fletcher and La Flesche 1972:272; Grinnell 1972:175-76, 187, 217-18, 225-27; McHugh 1972:93; Weltfish 1965:90, 369). While this list is far from exhaustive, it illustrates the range of nonfood uses for by-products from bison of each sex.

It is now commonplace for archeologists to investigate sex structure in prehistoric bison kills. Such information provides important insights into procurement strategies; and, in conjunction with other data such as presence of fetuses and stage of dental eruption and wear, it has proved a useful indicator of the season when the hunt took place.

The sex structure of a kill population is usually de-termined by sexing skulls or, more commonly when these are poorly preserved, by sexing mandibles or one or more post-cranial elements, notably metapodials (e.g., Dibble and Lor-rain 1968; Reher 1974; Reher and Frison 1980; Bedord 1974, 1978). Most prehistoric kill sites from which bison remains have been sexed have turned out to contain primarily females (e.g., Dibble and Lorrain 1968; Frison 1973, 1974, 1978; but for examples with high proportions of males see Frison, Wil-son, and Wilson 1976 and Wheat 1972). These results are not unreasonable, since most of these same sites, on the basis of tooth eruption and wear, have been found to be fall and/ or winter kills, a time of year when bison cows were in comparatively good condition and were most readily manipulated.

While the majority of these kill sites may in fact contain primarily females, most studies of the sex structure of kill populations have implicitly assumed that the proportions of male and female skulls, mandibles, or metapodials recovered archeologically directly reflect the proportions of males and females killed. This requires that cows and bulls, once killed, invariably be subjected to nearly identical processing. In light of frequent references in the historic and ethnographic literature to differential treatment of male and female bison, this assumption is untenable.

The present study examines in detail one clear-cut case of both selective procurement and selective processing, in which an initial kill population dominated by bulls was transformed by preferential removal of male elements into a discarded residue overwhelmingly dominated by female elements. The specific case to be considered is the late prehistoric spring-season Garnsey Bison Kill in the Pecos Valley of southeastern New Mexico.

The study begins with a general overview of the site, summarizing its stratigraphy and chronology, the method and seasonal timing of procurement, and the age and sex structure of the original kill population. It then turns to a

discussion of the nature of processing at Garnsey, examining
in detail the evidence for selectivity by the hunters as
reflected in the proportional frequencies of various male
and female skeletal parts abandoned at the site. The observed frequencies are evaluated in terms of the elements'
probable utility values (Binford 1978), providing insights
into some of the major factors that guided or constrained
procurement and processing strategies at the site. The
study concludes with a discussion of both methodological and
more general implications of the Garnsey case for understanding late prehistoric bison procurement in the Southern
Plains.

2. The Garnsey Bison Kill

INTRODUCTION

Detailed reports on two seasons of fieldwork at the
Garnsey Site (1977 and 1978) have been published elsewhere
(Speth and Parry 1978, 1980). Here I will provide only a
general overview of the site. Readers interested in more
specific information concerning the kill, and the recovered
faunal and archeological materials, should consult the site
reports.

PHYSICAL SETTING

The Garnsey Bison Kill (LA-18399) is on the eastern
edge of the Pecos River Valley about 20 km (12 mi) southeast
of Roswell, Chaves County, New Mexico (figs. 1-6). Bison
remains are exposed in the walls of a modern arroyo that is
actively cutting into and destroying the alluvial fill of a
broad, shallow wash (Garnsey Wash). The wash is one of many
similar dry washes that drain westward into the Pecos from a
low divide at the Llano Estacado or Caprock.

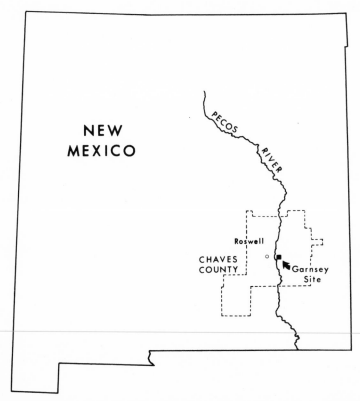

Fig. 1. Location of Garnsey Site in southeastern New Mexico.

The terrain between the Llano and the Pecos, known as
the Mescalero Pediment, consists of low, rolling plains dot-
ted with playas and in many places covered with extensive
dunes. The east side of the Pecos Valley in the Roswell
area is bounded by sheer cliffs, some nearly 30 m high.
Along the base of the cliffs, and cutting deeply into them,
are numerous large circular sinkholes, in several cases

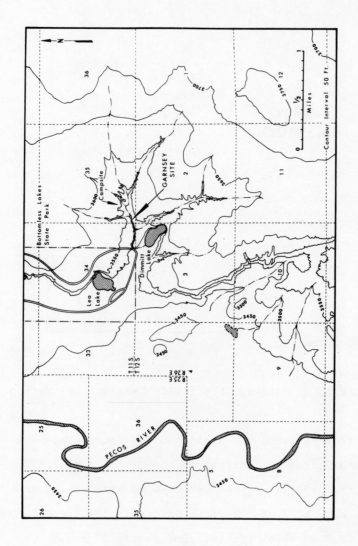

Fig. 2. Bottomless Lakes area, New Mexico, showing location of Garnsey Site (adapted from USGS Bottomless Lakes 7.5 Minute Quadrangle, Topographic Series).

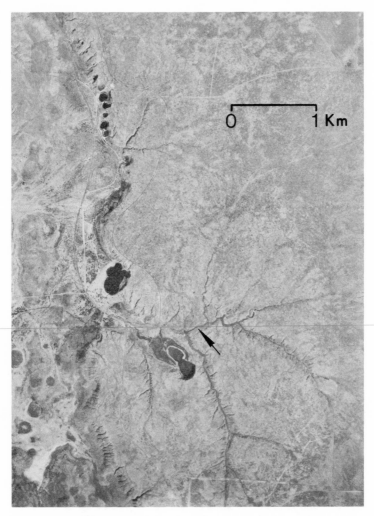

Fig. 3. Aerial photograph of Bottomless Lakes area
(arrow denotes location of Garnsey Bison Kill Site)
(BLM-ARS Series; date 09-30-73; image 7-51-21;
original negative scale 1:31, 680).

Fig. 4. Aerial view of Garnsey Wash facing west (left arrow indicates location of principal excavations at kill
site; center arrow denotes location of spring; right arrow indicates position of test trenches at Garnsey
Spring campsite) (Dimmitt Lake and Lea Lake visible in
background; photo taken July 1978).

forming deep lakes. Six of the largest lakes are included

within Bottomless Lakes State Park. A seventh, privately

owned Dimmitt Lake, lies just south of the park (fig. 2).

The mouth of the Garnsey Wash opens into the Pecos Val-

ley near the southeastern boundary of the park, between Dim-

mitt Lake and Lea Lake. The site itself is about 350 to 650

m upstream or east from the mouth (figs. 3 and 4).

The wash today is dry throughout most of the year.

Temporary ponded water occurs in some of the more deeply

Fig. 5. Aerial view of Garnsey Wash facing south, showing
principal excavations at kill site in 1977 (near side of
arroyo to right of van; backfilled) and in 1978 (both
sides of arroyo opposite van and at extreme right) (East
Dimmitt Wash and vertical sink wall of Dimmitt Lake visible in right background; photo taken July 1978).

scoured stretches of the arroyo after heavy spring rains,

and occasionally the arroyo flows for several hours after

summer cloudbursts. Permanent potable water is available at

nearby Dimmitt Lake and also at a spring in a small tribu-

tary off the main wash about 300 m east (upstream) of the

site (fig. 4). The water from the spring supports a rela-

tively lush growth of grasses (principally sacaton, Sporobo-

Fig. 6. General view of Garnsey Wash and modern arroyo,
facing southwest (south wall of Dimmitt sinkhole visible in background; datum C at instrument in left
foreground; photo taken June 1977).

lus airoides) and trees (historically introduced salt cedar,

Tamarix pentandra). The water rapidly seeps into the ground

or is lost to evaporation, so that none reaches the floor of

the main wash.

The present-day vegetation in the area of the site has

been altered significantly by arroyo cutting and subsequent

lowering of the water table. Overgrazing also has helped

reduce the local vegetation cover and has favored increases

in more drought-resistant species. Today the floor of the wash supports a relatively sparse growth of sacaton, black grama grass (Bouteloua eriopoda), broom snakeweed (Gutierrezia sarothrae), prickly pear (Opuntia spp.), yucca (Yucca spp.), and honey mesquite (Prosopis juliflora). Salt cedar occurs in the arroyo itself and near the spring. On the slopes above the wash, dominant grasses are black grama and tobosa (Hilaria mutica). Prickly pear and yucca occur sporadically. Shrubs, infrequent near the site, include mesquite, apache plume (Fallugia paradoxa), four-wing saltbush (Atriplex canescens), and several others (fig. 6).

PRESENT-DAY CLIMATE

The climate of the area today is semiarid. Roswell receives approximately 295 mm (11.62 in) of rain annually (Houghton 1974:802). Winters are relatively dry, with the entire time from November through March accounting for only 58 mm (2.28 in). Therefore over 80% (237 mm or 9.34 in) of the average annual precipitation occurs during the seven months from April through October. The rains come in two distinct periods, the first in May followed by a slight decline in June. The principal rainy season occurs in July, August, and September. These three months average approximately 132 mm (5.21 in), or nearly 45% of the annual rainfall. These late-summer rains generally come in the form of afternoon and evening thunderstorms, noteworthy for their

great variability in intensity, frequency, and spatial occurrence. Hail may also occur.

The average annual precipitation values in southeastern
New Mexico and the Southern High Plains do not differ greatly from values recorded in many parts of the Southwest or in
the Central and Northern High Plains (cf. Court 1974:
fig. 11). Looked at more closely, however, there are certain aspects of the precipitation pattern in the Southern
High Plains that differ from those in surrounding areas and
that may have been very important to human groups utilizing
the region. First, a lower proportion of the total annual
precipitation falls during the winter months in southeastern
New Mexico than in other areas of the Southwest, and peak
summer rainfall comes somewhat later in the season. Second,
the average ratio of median precipitation to mean precipitation in the Southern High Plains (less than 75%) more closely resembles values in southern Arizona and the Great Basin
than values on the Colorado Plateau or in the Central or
Northern Plains (Court 1974:212). Finally, the coefficient
of variation of annual precipitation for the period 1931-60
(over 40%) is higher than anywhere else in the Southwest or
Plains (Court 1974:211).

Summers in Roswell are warm. Average daily maxima ex-
ceed 32°C (90°F) from June through August. Daytime highs
above 38°C (100°F) are common. Winters are mild. The average daily high, even in January, the coldest month, is

12.8°C (55.1°F). Average minimum daily temperatures drop
below freezing from mid-November through mid-March, but
temperatures rarely reach 0°F (-18°C). The average length
of the frost-free season is 206 days, from 7 April through
30 October (Von Eschen 1961:51).

EXCAVATION

Two seasons of excavation were conducted at the Garnsey
Site (1977 and 1978) under Federal Antiquities Act permit
77-NM-037, issued to the University of Michigan by the
United States Department of the Interior and administered by
the Bureau of Land Management in the State of New Mexico
(Roswell District). Two excavation units were opened in
1977 (77-1 and 77-2; see fig. 7), exposing a total area of
approximately 120 m^2. In 1978 the unexcavated strip between
77-1 and the northern edge of the arroyo was examined
(trench 78-1). This unit opened an area of about 44 m^2. In
addition, a small unit, 8 m^2, was opened at the northern
edge of 77-1 (78-5). Five trenches were opened on the south
side of the arroyo in 1978 (78-2, 60 m^2; 78-3, 38 m^2; 78-4,
60 m^2; 78-6, nearly 8 m^2; 78-7, 9 m^2), bringing the total
area excavated in the second season to 227 m^2.

The two southernmost rows of grid squares in trench
78-2 were terminated after preliminary testing indicated
they were sterile. Trench 78-3, a long, narrow unit extend-

ing south away from the arroyo, also proved sterile and was
terminated.

Trenches 78-6 and 78-7 were small units exposed on the
south side of the arroyo nearly 100 m downstream (west) from
78-2. These units were opened to salvage large quantities
of bison bone unexpectedly exposed early in the 1978 season
by flash flooding. Trench 78-6 was the first of these units
to be opened and was begun before the grid system had been
extended to this area of the site. The orientation of the
unit therefore deviates from that of the other trenches.

STRATIGRAPHY

Bison remains are exposed in both walls of the Garnsey
arroyo along a stretch of nearly 300 m, beginning approximately 350 m upstream from the mouth of the wash. The remains form numerous small, spatially discrete clusters, each
containing up to several hundred almost completely disarticulated bones. These clusters vary in depth from about 1.2
to 1.7 m below the present surface of the wash, although
scattered remains and occasional small clusters occur both
at shallower and at greater depths. Nowhere in the Garnsey
Site do bison remains form thick, continuous layers comparable to the "bone beds" of many Northern Plains kills
(e.g., Vore Site; see Reher and Frison 1980).

Four stratigraphic units (A through D) were recognized
in the 2- to 4-m thick section of alluvium exposed by the

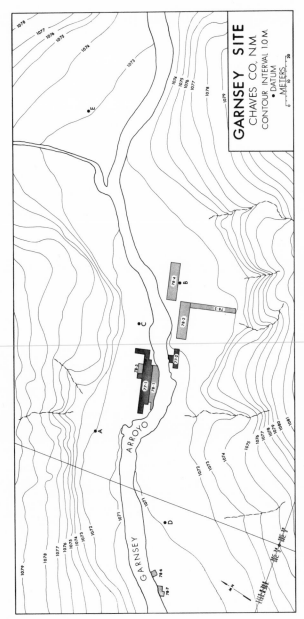

Fig. 7. Garnsey Site (LA-18399), showing areas excavated in 1977 and 1978.

Garnsey arroyo. Only the upper three (A through C) are of concern here. Unit A, the uppermost unit, consists of approximately 1.2 m of finely laminated silt and sand with occasional thin lenses and stringers of fine gravel, particularly in the lower part of the unit (fig. 8; sediment samples V-ZZZ).

Unit B, the principal archeological unit at the site, is considerably coarser than unit A, consisting of numerous lenses of pea-sized and larger gravel (fig. 8; sediment samples K-U). The gravel lenses are distinctly graded, and in many places several graded sequences are directly superposed. Within unit B is a distinctive greenish gray cienega or "wet-meadow" horizon that may be traced along the walls of the arroyo for more than 100 m (fig. 8; sediment sample P). In places the horizon is only 10-15 cm thick, but elsewhere it exceeds half a meter. When wet the cienega displays blotchy, rust-colored limonite mottling. When dry it becomes extremely hard, with distinct vertical blocky jointing. Although the clay content of the cienega is much higher than in the rest of unit B, the presence of thin lenses of gravel within the horizon attests to continued periodic torrential deposition.

Unit C, 2.5 to 4.0 m below the surface of the wash, consists of deposits somewhat finer than those of unit A and lacks the fine laminae that characterize the latter (fig. 8; sediment samples C-J). A thick greenish gray cienega hori-

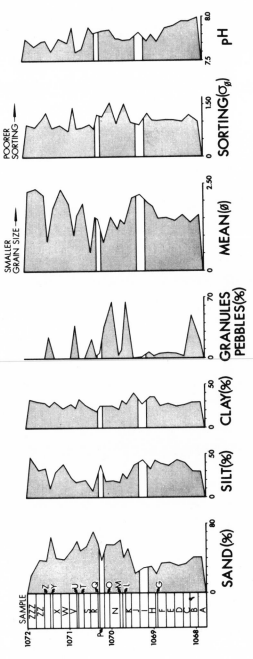

Fig. 8. Textural and pH analysis of deep arroyo profile (south edge of grid square H-2 or 508S502W; see figs. 7 and 10 for approximate location), showing stratigraphic units A-D and principal cienega horizons (sediment samples I and P). Numbers at left of figure give absolute elevation in meters.

zon, very similar to the one in unit B, occurs near the top of unit C (fig. 8; sediment sample I).

The cienega horizons in units B and C can be followed along the walls of the arroyo for considerable distances. Thus they provide useful horizon markers in deposits that otherwise are characterized by rapid and frequent lateral facies changes. In many parts of the arroyo the cienegas are visible simultaneously in both walls. Elsewhere they are present in only one wall, where they may be traced for several meters before shifting to the opposite wall. When followed laterally out of the main arroyo into minor tributaries, the horizons rise slightly and pinch out within a few meters. The cienegas therefore appear to delineate the course of a major broad channel, probably on the order of 5 m wide and perhaps up to 1 m deep, that may have been perennially moist or have contained ponded water. Periodic torrential flooding deposited lenses of gravel within these moist or ponded stretches of the wash. The consistent superposition of cienega horizons indicates that the location of the main channel has remained relatively stable throughout unit C and unit B times. Moreover, the modern arroyo, which is cutting into the alluvial deposits in the wash, appears to be following more or less the same course.

During unit B and unit C times, there were two or three small, shallow, braided channels paralleling the main channel on the north. These peripheral channels contain graded deposits of gravel but show no evidence of having contained ponded water. During unit A times the peripheral channels were filled in with fine sediments, confining flow to the main channel.

The vast majority of bison remains and associated cultural material are found within the upper half-meter of unit B, above the cienega horizon. Occasional bison remains occurred in the lower part of unit A, and a few were found within or below the unit B cienega. In all, six thin, spatially discontinuous "bone levels" were recognized (A1-2; B3-6). Of these, only level B3 contained significant quantities of material (nearly 90% of the total inventory of bones from the two seasons of excavation).

SPATIAL DISTRIBUTION OF THE BISON REMAINS

Regardless of stratigraphic level, the bison remains are confined to a relatively narrow zone along both flanks of the main channel and along the flanks and bottoms of the small parallel channels. Originally the remains probably also extended across the bottom of the main channel, but these have been obliterated in the past twenty to thirty years by arroyo cutting. Although the overall distributional pattern of bison remains is to some extent a product of periodic torrential flooding in the Garnsey Wash, it largely reflects a procurement strategy in which bison were sur-

rounded, or ambushed, and butchered in or close to the main
channel.

Figure 9 is a computer-generated north-south section or
"backplot" that illustrates the tendency at Garnsey for
bones to be concentrated along the flanks and bottoms of
channels. The plot was constructed by projecting all the
bones in eight parallel north-south rows of grid squares at
the east end of the site (trenches 77-1, 78-1, and 78-5)
onto a single vertical north-south plane, so that only the
depth of each item and its north-south position are considered. The figure shows a concentration of bones, particularly in level B3, on the north flank of the main prehistoric channel, whose position more or less coincided with that
of the modern arroyo (i.e., to the right or south of square
511S). The plot also shows a heavy concentration of material on the flanks and bottom of a small, shallow channel to
the left or north of the main channel (i.e., centered approximately in squares 505S and 506S).

The bison remains in each "bone level" occur in small,
spatially discrete clusters or concentrations, separated
from similar quasi-contemporary clusters by zones with few
or no remains (see figs. 10, 11, and 12 for the spatial distribution of level B3 materials; the sample size from other
levels is too small to warrant illustration). While most
clusters probably reflect discrete procurement events, the
time between events within the same level is probably very

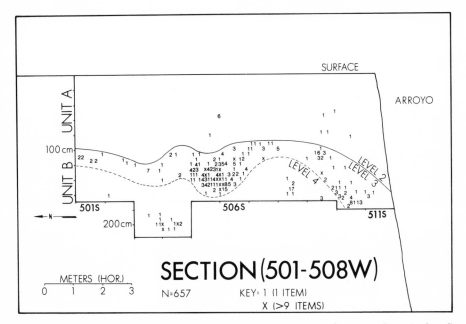

Fig. 9. Computer-generated north-south section or "backplot" (viewer facing east) of bison bones on north side of modern arroyo; plot constructed by projecting all items in eight parallel north-south rows of grid squares (501W to 508W; trenches 77-1, 78-1, and 78-5) onto a single north-south plane.

short, perhaps only a few days or weeks or at most a few years. Radiocarbon dating (see below) indicates that the entire 2- to 4-m alluvial sequence exposed in the Garnsey arroyo represents a comparatively short period, probably on the order of a few centuries or less.

Individual kill events varied in size from relatively small ones, in which only four or five animals were taken, to occasional substantial ones, in which fifteen to twenty

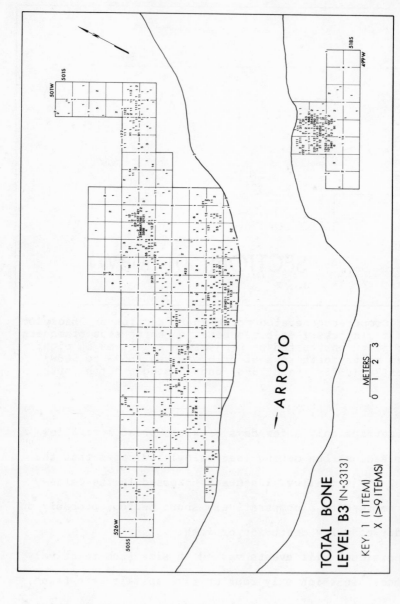

Fig. 10. Computer-generated scatterplot of level B3 bison remains in trenches
77-1, 77-2, 78-1, and 78-5 (see fig. 7 for trench designations).

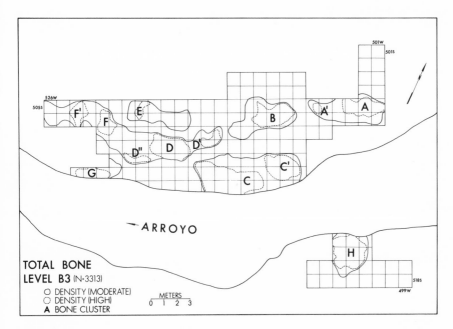

Fig. 11. Principal level B3 bone clusters in trenches 77-1, 77-2, 78-1, and 78-5 (see fig. 7 for trench designations).

or more animals were killed (the largest cluster, which was partially destroyed by arroyo cutting and only partially excavated, contained at least fifteen skulls). A "typical" or average kill probably involved six to eight animals.

Butchering reduced the bison carcasses to piles of almost totally disarticulated remains. To some extent the disarticulation is due to fluvial disturbance (to be discussed more fully below), but even in clusters where such disturbance was minimal, very few elements were found in anatomical order. Articulated parts were most commonly the

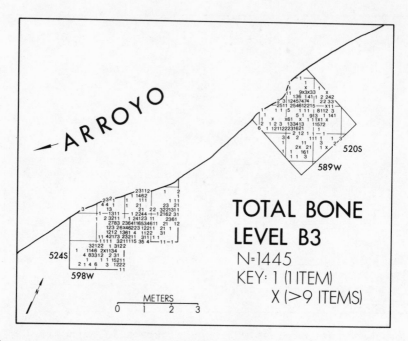

Fig. 12. Computer-generated scatterplot of level
B3 bison remains in trenches 78-6 and 78-7 (see
fig. 7 for trench designations).

feet and lower portions of the forelimbs (from the radius
down) or occasional short series of thoracics, lumbars, or
cervicals. Butchering units such as those encountered at
Olsen-Chubbuck (Wheat 1972) or Casper (Frison 1974) were
virtually nonexistent (the single exception was the front
end of an adult female that included the head, neck, forelimbs, and anterior portion of the thorax).

Limited secondary processing took place near the edge
of the main channel somewhat upstream from the kill-

butchering areas (trenches 78-2 and 78-4; see fig. 7). Two small basin-shaped hearths were found in level B3 in this area, ringed by aprons of highly fragmented bone (fig. 13), hundreds of tiny retouch or resharpening flakes, and a variety of formal stone tools. A north-south section or "backplot" through the processing area (fig. 14) illustrates the stratigraphy in this portion of the site and clearly shows the location of the south flank of the main prehistoric channel. Most of the bone fragments were unidentifiable, though many clearly were derived from limbs, ribs, and vertebral processes (table 1). Fire-cracked rock and ceramics were extremely rare.

It is quite likely that similar small processing areas remain undiscovered within the wash, each situated in the general vicinity of one of the kill-butchering localities. Although a large quasi-contemporary campsite has been located on a terrace above the wash about 400-500 m east of Garnsey (figs. 2 and 4), there is no evidence that activities at this site were connected to those at the kill. No other campsites, which might have served as processing localities associated with Garnsey, have been found in the area.

DESCRIPTION AND SPATIAL
DISTRIBUTION OF THE ARTIFACTS

The sample of artifacts from the Garnsey Site is relatively small and is overwhelmingly dominated by unmodified

Table 1

Inventory of level B3 bison remains from processing
area on south side of arroyo (trenches 78-2 and 78-4)

Element	N	%	Element	N	%
1. Skull	0	0.0	27. Ulnar carpal	0	0.0
2. Mandible	2	0.1	28. Accessory carpal	0	0.0
3. Hyoid	0	0.0	29. Fused 2d-3d carpal	1	0.1
4. Incisor/canine	4	0.3	30. 4th carpal	0	0.0
5. Molar/premolar	2	0.1	31. 5th metacarpal	1	0.1
6. Misc. tooth fragments	23	1.6	32. Metacarpal	5	0.4
7. Atlas	0	0.0	33. Pelvis	4	0.3
8. Axis	0	0.0	34. Femur	2	0.1
9. Cervical (3-7)	0	0.0	35. Patella	0	0.0
10. Thoracic (1-14)	4	0.3	36. Tibia	1	0.1
11. Lumbar (1-5)	1	0.1	37. Astragalus	2	0.0
12. Sacrum	0	0.0	38. Lateral malleolus	0	0.0
13. Caudal	5	0.4	39. Calcaneus	1	0.1
14. Unident. vertebral body	1	0.1	40. Naviculocuboid	0	0.0
15. Unident. vertebral pad	2	0.1	41. 1st tarsal	0	0.0
16. Unfused vertebral summit	0	0.0	42. Fused 2d-3d tarsal	0	0.0
17. Rib	26	1.9	43. 2d metatarsal	0	0.0
18. Rib-vertebral process fragments	140	10.0	44. Metatarsal	3	0.2
19. Costal cartilage	0	0.0	45. Metapodial	1	0.1
20. Sternebra	2	0.1	46. 1st phalanx	7	0.5
21. Scapula	1	0.1	47. 2d phalanx	4	0.3
22. Humerus	1	0.1	48. 3d phalanx	1	0.1
23. Radius	1	0.1	49. Proximal sesamoid	4	0.3
24. Ulna	0	0.0	50. Distal sesamoid	1	0.1
25. Radial carpal	0	0.0	51. Unident. bone fragments	1,140	81.7
26. Intermediate carpal	2	0.1	Total	1,396	100.2

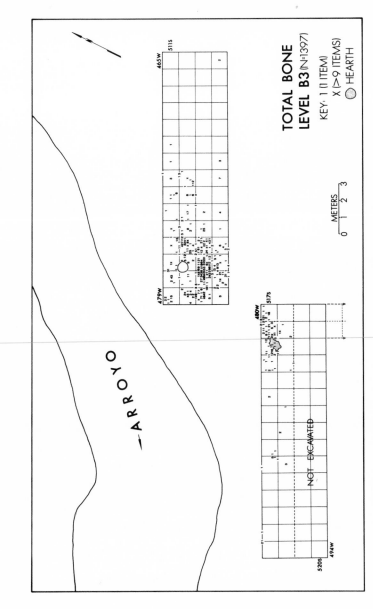

Fig. 13. Computer-generated scatterplot of level B3 bison remains in the processing area (trenches 78-2 and 78-4; see fig. 7 for trench designations).

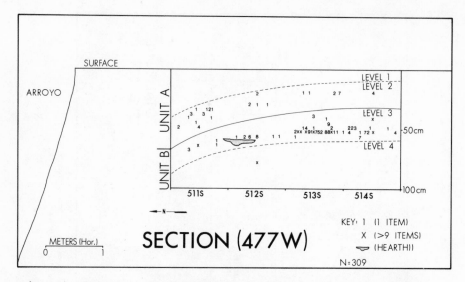

Fig. 14. Computer-generated north-south section or "back-plot" (viewer facing east) of bison bones in processing area on south side of modern arroyo; plot constructed by projecting all items from single north-south row of grid squares (477W; trench 78-4) onto a single north-south plane.

flakes and tiny retouch and resharpening spalls. Of a total of 801 lithic items, 459 (57.3%) are small, unmodified flakes, 253 (31.6%) are tiny retouch and resharpening spalls, most of which come from the secondary processing area, and 40 (5.0%) are larger utilized flakes. Only 49 (6.1%) formal lithic tools were recovered. Of these, 3 (0.4%) are cores, 19 (2.4%) are unifacially retouched pieces (excluding endscrapers), 6 (0.7%) are endscrapers, 3 (0.4%) are ovate bifaces, 8 (1.0%) are bifacial knife fragments, and 10 (1.2%) are projectile points. Several broken limb bones may have served as choppers, but none are unambiguous tools.

The spatial distribution of level B3 lithic tools, re-sharpening flakes, and debitage in the kill-butchering and processing areas is illustrated in figures 15, 16, and 17. The boundaries of the lithic clusters correspond very close-ly to those of the bison bones. The relative abundance of artifacts in the processing area, particularly projectile points and scrapers, compared with the scarcity of these items in the bone clusters, is noteworthy.

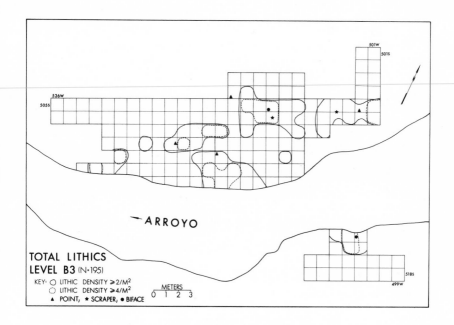

Fig. 15. Computer-generated scatterplot of level B3 lithic items in trenches 77-1, 77-2, 78-1, and 78-5 (see fig. 7 for trench designations).

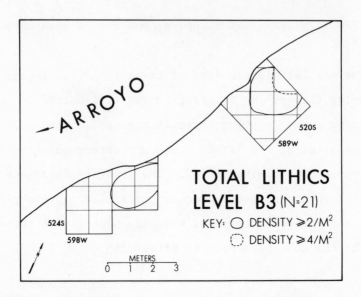

Fig. 16. Computer-generated scatterplot of level
B3 lithic items in trenches 78-6 and 78-7 (see
fig. 7 for trench designations).

In the second field report on the Garnsey Site (Speth
and Parry 1980:165-67), it was observed that the values of
most size-related attributes of unmodified flakes and debitage (e.g., length, width, weight) decreased progressively
from level B4 to level A2, while the values of several technological attributes (e.g., medial axis angle, length/width
ratio, and perhaps platform angle) remained nearly constant
or showed no clear-cut trend. This apparent directional
trend in flake size was dismissed as an artifact of our having included in the computations a very large and heavy outlier in level B4 and a moderately large outlier in level B3.

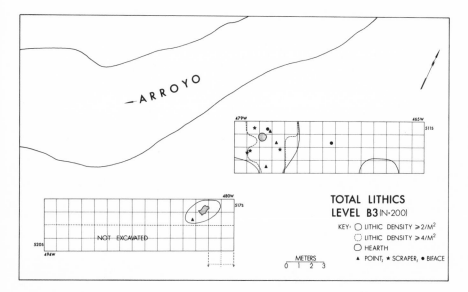

Fig. 17. Computer-generated scatterplot of level B3
lithic items in processing area (trenches 78-2 and
78-4; see fig. 7 for trench designations).

Subsequent reexamination of these data using median rather
than mean values revealed that the trend was indeed real
(the median is much less sensitive than the mean to the in-
clusion of extreme values). Finally, when four exception-
ally large outliers were excluded, the trend was enhanced
rather than diminished. The data (minus these four outliers) are summarized in table 2.

Experimental fracture studies in which steel balls were
dropped from an electromagnet onto glass prisms of different
sizes and platform angles have shown that flake size decreases with decreasing core (or tool) size, all other

Table 2

Metric attributes of unmodified flakes (≥4 mm)
by stratigraphic level from Garnsey Site

Level/Attribute	N	Minimum	Maximum	Mean	Median	S.D.
Level A2						
Length (mm)	74	2.0	58.0	10.7	8.5	9.2
Width (mm)	74	3.0	37.0	9.6	8.0	6.2
Bulb thickness (mm)	51	0.5	10.0	2.0	1.5	1.6
Maximum thickness (mm)	74	0.5	16.0	2.3	1.0	2.6
Medial axis angle (°)	46	70.0	140.0	96.1	95.0	13.1
Platform angle (°)	43	20.0	125.0	64.2	60.0	20.8
Weight (g)	74	0.0	24.2	0.9	0.1	3.3
Platform width (mm)	49	1.0	21.0	5.3	4.0	4.2
Platform thickness (mm)	49	0.0	7.0	1.6	1.0	1.3
Length/width ratio	74	0.2	3.4	1.2	1.1	0.6
Level B3						
Length (mm)	419	2.5	65.5	12.2	10.0	8.2
Width (mm)	419	2.5	42.0	11.1	9.5	6.4
Bulb thickness (mm)	207	0.5	16.0	2.7	2.0	2.1
Maximum thickness (mm)	419	0.5	19.0	3.0	2.0	2.5
Medial axis angle (°)	179	55.0	155.0	92.6	90.0	13.1
Platform angle (°)	163	10.0	135.0	69.0	70.0	18.4
Weight (g)	419	0.0	24.7	1.0	0.2	2.9
Platform width (mm)	195	0.5	30.0	6.1	5.0	4.5
Platform thickness (mm)	197	0.0	16.0	2.0	1.5	1.9
Length/width ratio	419	0.3	3.7	1.2	1.1	0.5
Level B4						
Length (mm)	87	1.1	42.0	14.0	12.0	8.0
Width (mm)	87	3.0	55.0	12.8	9.5	8.7
Bulb thickness (mm)	44	0.5	17.5	3.6	2.5	3.8
Maximum thickness (mm)	87	1.0	18.0	3.7	3.0	3.5
Medial axis angle (°)	41	55.0	125.0	97.3	100.0	15.7
Platform angle (°)	35	30.0	115.0	68.9	70.0	19.6
Weight (g)	87	0.0	14.2	1.3	0.3	2.8
Platform width (mm)	42	1.0	27.0	7.3	5.5	5.7
Platform thickness (mm)	40	0.0	17.0	2.6	2.0	2.9
Length/width ratio	87	0.2	2.5	1.2	1.1	0.5

parameters being equal (i.e., platform angle, drop height, and ball diameter; Speth 1981; see also Dibble and Whittaker 1981; Raab, Cande, and Stahle 1979:176). The implication of these experimental results for the Garnsey data is that the average size of the tools being flaked or resharpened progressively decreased from the older to the younger levels. Obviously, several possible scenarios might account for this trend. The one that is tentatively favored here, for reasons that will become more apparent later in the discussion, is that the amount of on-site processing increased in the younger levels in response to increasing transport constraints. Given the absence of suitable chert in the vicinity of Garnsey, this would have necessitated longer use of the tools brought to the site, which in turn would have led to greater reduction in average tool size through increased attrition and more frequent resharpening.

Eight of the ten projectile points from Garnsey are small triangular side-notched varieties, only two of which are complete or nearly complete (figs. 18 and 19). Of those with bases sufficiently well preserved to permit identification, two have basal notching (Harrell type) and three lack basal notching (Washita type; cf. Bell 1958). Among the latter, two have concave bases and one has a nearly straight base.

Of the remaining two points, one is an unidentifiable, lightly serrated tip or blade fragment, the other is a large

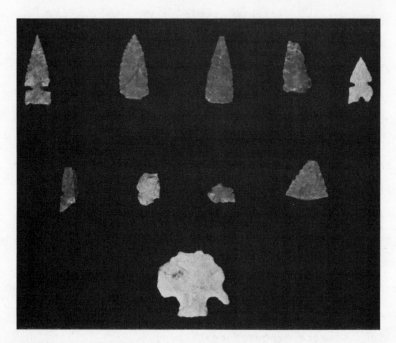

Fig. 18. Photographs of projectile points from Garnsey
Site. Top row, from left to right: H-17/1, D-8/1, E-2/4,
D-7/5, 511S477W/14; middle row, from left to right:
518S483W/2, 512S476W/29, 514S477W/S(7/8/78), D-12/1; bottom row: I-13/1.

barbed, corner-notched point that has been broken and bifa-

cially reworked into a hafted knife (figs. 18 and 19).

LITHIC MATERIALS

Lithic raw materials recovered at Garnsey are highly

varied in color and texture. A total of twenty-two types

was recognized (table 3), although several of these may ac-

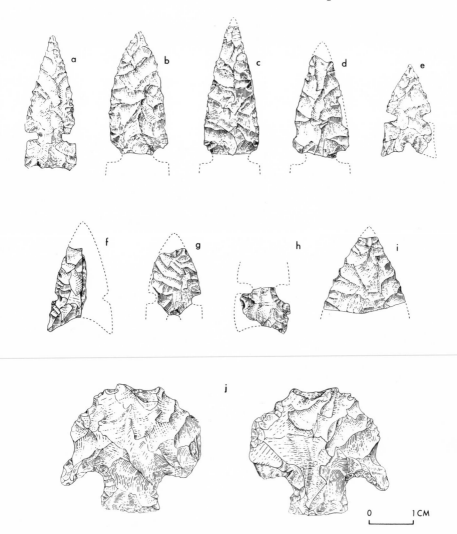

Fig. 19. Illustrations of projectile points from Garnsey
Site. (a) H-17/1; (b) D-8/1; (c) E-2/4; (d) D-7/5; (e)
511S477W/14; (f) 518S483W/2; (g) 512S476W/29; (h)
514S477W/S(7/8/78); (i) D-12/1; (j) I-13/1.   (Drawn by
Margaret Van Bolt, University of Michigan.)

tually be variants or facies of the same material. The
sources for many of the types are uncertain. Several may
have been procured as cobbles from gravels in the Pecos Valley. While no systematic investigations of these gravels
have as yet been undertaken, preliminary surveys indicate
that suitable materials are not available within a radius of
several kilometers of the site. Bedrock in the immediate
vicinity of the site (gypsum) is extremely soft and essentially useless for the manufacture of tools, including hammers or choppers. The most abundant lithic materials at the
site, gray and mottled gray cherts (types 11 and 12 in table
3), probably derive from limestone ridges (Permian San An-
dres Formation) west of Garnsey between the Pecos Valley and
the Sacramento Mountains (Kelley 1971). The nearest chert-
bearing exposures are about 25-30 km west of the site, but
similar materials would also have been available from localities as much as 100-140 km away. Four small pieces of ob-
sidian were recovered; the nearest sources for these are
several hundred kilometers to the northwest, west, or southwest. Seventeen pieces of Tecovas and Alibates chert are
the only materials demonstrably derived from sources to the
east of Garnsey. Both types come from the Texas Panhandle,
a minimum of 300-400 km to the east and northeast. Of the
projectile points, 70% (seven of ten) are made of gray or
mottled gray chert from sources west of the site (types 11
and 12); none are made of eastern materials.

Table 3

Lithic raw materials by stratigraphic level utilized at Garnsey Site

Type	Description	Level B4 N	Level B4 %	Level B3 N	Level B3 %	Level A2 N	Level A2 %
1	Obsidian	0	0.0	3	0.7	1	1.4
2	Glassy black basalt	0	0.0	1	0.2	1	1.4
3	Fine-grained igneous	1	1.1	10	2.3	0	0.0
4	Unidentified, weathered, nonsiliceous	0	0.0	8	1.8	0	0.0
5	Quartzite (light to dark gray, maroon)	6	6.7	11	2.5	1	1.4
6	Mottled, milky white chalcedony	11	12.2	18	4.1	11	15.1
7	Gray chalcedony	6	6.7	12	2.7	2	2.7
8	Yellow, orange, tan, or brown chalcedony	5	5.6	7	1.6	0	0.0
9	Banded olive, brown, tan, and gray chert	1	1.1	5	1.1	0	0.0
10	Grainy pink chalcedony	0	0.0	1	0.2	0	0.0
11	Mottled, predominantly gray chert	26	28.9	169	38.3	27	37.0
12	Gray chert	11	12.2	102	23.1	27	37.0
13	Green or olive chert	1	1.1	6	1.4	0	0.0
14	Alibates, Tecovas, and related materials	2	2.2	15	3.4	0	0.0
15	White chert	1	1.1	1	0.2	0	0.0
16	"Fingerprint" chert	3	3.3	8	1.8	1	1.4
17	Mottled black chert or chalcedony	5	5.6	16	3.6	0	0.0
18	Dark brown chert or chalcedony	1	1.1	4	0.9	0	0.0
19	Chalky white, tan, or light gray chert	4	4.4	26	5.9	1	1.4
20	Calichelike (burned?)	6	6.7	11	2.5	1	1.4
21	Yellow brown jasper	0	0.0	2	0.5	0	0.0
22	Oolitic chert	0	0.0	3	0.7	0	0.0
	Total	90	100.0	439	100.0	73	100.2

Several interesting changes occur in the use of lithic raw materials at Garnsey during the time represented by levels B4, B3, and A2. The following are particularly noteworthy: (1) the total number of different material types used at the site decreases from sixteen in level B4 to ten in level A2 (table 4); (2) there is a clear progressive increase from unit B to unit A in the value of the coefficient of variation (C.V.) for the number of flakes per material type (table 4); (3) the proportion of gray and mottled gray chert and chalcedony (types 7, 11, and 12) increases from less than 50% in level B4 to nearly 90% in level A2 (table 3); and (4) all of the demonstrably eastern lithic materials (Tecovas and Alibates; type 14) occur in unit B (table 3).

These observations suggest that hunters, in preparing for the kills in the younger levels at the site, had access to a restricted range of lithic materials available primarily on the limestone ridges west of the Pecos Valley. Sources within and east of the valley, which had been important to hunters in the earlier levels, had dropped out almost entirely by unit A times.

Considered in conjunction with the evidence noted above for increased on-site processing in the younger levels (i.e., as reflected by decreasing overall size of retouch and resharpening flakes), the lithic data very tentatively suggest that during the course of the fifteenth century hunters were coming to Garnsey from spring-season settle-

ments that may have been shifting farther and farther to the
west of the Pecos. Moreover, since the changes in the lithic data are gradational rather than abrupt, it appears that
we are looking at a gradual change in settlement location
within essentially the same spring-season procurement system
rather than the replacement of one system by another. Obviously, until a great deal more research specifically ad-
dressing these questions is undertaken at other settlements
in the region, these conclusions must remain little more
than speculation.

Table 4

Number of lithic raw material types per level and number
of flakes per raw material type at Garnsey Site

Unit/ Level	Number of Flakes	Number of Material Types	Number of Flakes per Material Type		
			Mean	S.D.	C.V.(%)[1]
A2	73	10	3.32	7.82	235.8
B3	439	22	19.96	38.41	192.5
B4	90	16	4.09	5.80	141.6

[1]C.V., Coefficient of variation (C.V. = 100 x S.D./mean).

EDGE ANGLES

Edge angles were determined on utilized flakes, unifacially and bifacially retouched tools, and resharpening
flakes with remnants of the original tool edge preserved on
their striking platforms. On tools with more than a single
working edge, each edge was treated separately, providing a

total sample size of 134 (table 5). Three clear-cut edge-
angle modes were observed (fig. 20). The first, or sharp
edge-angle mode (mean 21.0°; range 15°-25°), contains only
utilized flakes; none of the retouched edges and none of the
edges preserved on the platforms of resharpening flakes fall
into the sharp mode. The majority of the utilized flakes
(72.7%) fall within the intermediate edge-angle mode (mean
46.7°; range 30°-65°). Only two utilized edges (4.5%) fall
within the steep mode (mean 77.7°; range 70°-90°).

Table 5

Descriptive statistics for edge angles of each major
tool type recognized in Garnsey lithic assemblage

Tool Type	Number of Edges	Min.	Max.	Mean	S.D.
Utilized flake	43	15	80	39.0	14.5
Unifacially retouched (excl. endscraper)	19	35	90	60.8	16.9
Endscraper	7	40	90	67.9	18.7
Projectile point	9	32	50	41.9	6.4
Ovate biface	6	30	80	65.0	20.5
Biface or knife	9	40	80	55.2	13.3
Resharpening flake	41	40	90	68.9	13.6

Edges with unifacial retouch are relatively evenly
divided between the intermediate and steep edge-angle modes
(43.2% and 56.8%). The spatial distribution of unifacially
retouched tools with intermediate edge angles differs markedly from that of similar tools with steep edges. In the
kill-butchering areas, more than 80% of these tools have

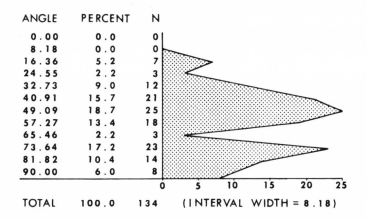

ANGLE	PERCENT	N
0.00	0.0	0
8.18	0.0	0
16.36	5.2	7
24.55	2.2	3
32.73	9.0	12
40.91	15.7	21
49.09	18.7	25
57.27	13.4	18
65.46	2.2	3
73.64	17.2	23
81.82	10.4	14
90.00	6.0	8
TOTAL	100.0	134

(INTERVAL WIDTH = 8.18)

Fig. 20. Frequency distribution of edge angles of all Garnsey tools (N = 134).

angles falling between 35° and 65°; whereas in the processing area, nearly 70% of these tools have angles greater than 75°.

Edges with bifacial retouch tend to be concentrated within the intermediate mode (62.8%), with a minority in the steep mode (37.2%). Virtually all the bifacial knife fragments were found in the processing area.

Edges preserved on the platforms of resharpening flakes fall largely within the steep edge-angle mode (65.0%); the rest have intermediate edge angles. This suggests that "resharpening" flakes are, to a large extent, debris from use-related attrition or reedging of tools with edge angles greater than about 70°.

Edge-angle distributions similar to the trimodal pattern observed at Garnsey have been noted in a variety of
other archeological assemblages. For example, Wilmsen
(1970:68-74) observed three comparable modes in Paleo-Indian
assemblages, and Frison (1970:36-38, 1974:92) documented
bimodal distributions in tools from two Wyoming bison kills
(Glenrock and Casper) comparable to the sharp and intermediate modes at Garnsey.

Both Wilmsen and Frison attribute the sharp edge-angle
mode to cutting. Frison (1970:36) suggests that tools with
intermediate edge angles were used in scraping, while Wilmsen (1970:70) posits a somewhat broader range of uses for
these tools, including both scraping and heavy-duty cutting.
Wilmsen (1970:71) attributes the steep mode to maintenance
tasks requiring tools with strong edges, such as wood working, bone working, or heavy shredding.

If Wilmsen's assessment of the general nature of tasks
performed with steep-edged tools is reasonable, their absence in primary kill sites such as Glenrock and Casper,
both of which lacked on-site processing areas, is not sur-
prising (Frison 1970:2, 1978:173). The presence of the
steep edge-angle mode at Garnsey, then, implies that the
range of activities was broader at this site than at either
Glenrock or Casper. This conclusion is supported by the
presence of at least one secondary processing area at Garnsey, characterized by hearths and moderate quantities of

highly fragmented limb bones and ribs. It is further supported by the fact that most unifacial tools with steep
edges were confined to the processing area. That other tool
types with steep edges were found within kill-butchering
areas, however, suggests that a relatively broad range of
processing activities may have been taking place throughout
the site. This would help account for the high degree of
disarticulation within the various bone clusters.

CHRONOLOGY

To determine the time span of bison hunting at the
Garnsey locality, an extensive program of radiocarbon dating
was undertaken (see Speth and Parry 1980:41 for a detailed
discussion of site chronology). Fourteen samples, all of
bone, were submitted to Geochron Laboratories, Cambridge,
Massachusetts, for analysis (table 6). Eleven of these were
from the major bone level (B3). Of the remaining samples,
one was from level B4, one from level B5 within the cienega
horizon, and one from level B6 below the cienega. Adequate
charcoal samples were not obtained during the excavations.
All bone dates were corrected for ^{13}C fractionation.

The fourteen samples provided twenty-one age estimates
(nine apatite, twelve gelatin). Of these, three came out
fully modern and were rejected. The remaining dates averaged 370±29 B.P. or A.D. 1580±29 (mean and standard devia-
tion weighted according to procedures outlined in Long and

Rippeteau 1974 and Davies 1961:133). This mean value yields
a MASCA-corrected age range (1 S.D.; new half-life of 5,730
years; Ralph, Michael, and Han 1973) of A.D. 1440-1520
(510-430 B.P.). The site is slightly younger, about A.D.
1502-98 (448-352 B.P.), if the age is calibrated using the
tables of Damon et al. (1974). A more recent calibration
scheme, developed by Stuiver (1982), yields an age estimate
more in line with the MASCA results (ca. A.D. 1445-1525 or
505-425 B.P.). No stratigraphic patterning is evident in
the dates; dates from within and below the cienega horizon
fall near the younger end of the range.

During the summer of 1980 a small sample of charred
grass culms (stems), apparently burned during a prehistoric
range fire, was collected from the margin of the main channel in unit A. This sample provided the first date for unit
A and the only nonbone date for the site. The material was
submitted to Beta Analytic, Inc., in Coral Gables, Florida.
The result, previously unpublished, is A.D. 1495±100
(455±100 B.P.; Beta-1925; ^{13}C fractionation-corrected; Libby
half-life of 5,568 years; age referenced to the year
A.D. 1950; date not MASCA-corrected).

Given the tendency for bone dates to be too young (Tamers and Pearson 1965), and the newly obtained date from unit
A, the "true" age of major procurement activities during
unit B times in the Garnsey Wash probably falls toward the
earlier end of the ranges given above, or about

Table 6

Radiocarbon dates from Garnsey Site

Provenience	Laboratory Number[1]	Unit/Level	¹³C Fractionation (per mil)	Age (±1 S.D.)[2]		Age (±1 S.D.)[3]	
				B.P.	A.D.	B.P.	A.D.
1. Ar/S/18	GX-5676-A[4]	B5	-0.6	290±145	1660±145	299±145	1651±145
2. Ar/S/18	GX-5676-G	B5	-8.3	285±130	1665±130	294±130	1656±130
3. Ar/S/19	GX-5677-A	B6	-1.3	310±125	1640±125	319±125	1631±125
4. Ar/S/19	GX-5677-G	B6	-8.4	145±120	1805±120	149±120	1801±120
5. Ar/N/33CD	GX-5678-A	B3	-0.7	155±120	1795±120	160±120	1790±120
6. Ar/N/33CD	GX-5678-G	B3	-8.8	ND[5]	ND[5]	ND[5]	ND[5]
7. 519S589W/16	GX-5679-A	B3	0.0	225±120	1725±120	232±120	1718±120
8. 519S589W/16	GX-5679-G	B3	-9.1	575±145	1375±145	592±145	1358±145
9. 519S590W/3	GX-5680-A	B3	+0.8	365±125	1585±125	376±125	1574±125
10. 519S590W/3	GX-5680-G	B3	-8.7	535±145	1415±145	551±145	1399±145
11. 519S590W/12	GX-5681-A	B3	-0.2	515±125	1435±125	530±125	1420±125
12. 519S590W/12	GX-5681-G	B3	-9.4	Modern[6]	Modern[6]	Modern[6]	Modern[6]
13. I-21/4	GX-5682-A	B3	-0.1	ND[5]	ND[5]	ND[5]	ND[5]
14. I-21/4	GX-5682-G	B3	ND[5]	ND[5]	ND[5]	ND[5]	ND[5]
15. J-12/7	GX-5683-A	B3	+0.1	445±105	1505±105	458±105	1492±105
16. J-12/7	GX-5683-G	B3	-8.5	345±100	1605±100	355±100	1595±100
17. J-13/4	GX-5684-A	B3	+0.3	1045±125	905±125	1076±125	874±125
18. J-13/4	GX-5684-G	B3	-10.7	ND[5]	ND[5]	ND[5]	ND[5]
19. J-23/1	GX-5685-A	B4	-0.4	ND[5]	ND[5]	ND[5]	ND[5]
20. J-23/1	GX-5685-G	B4	-8.7	Modern[6]	Modern[6]	Modern[6]	Modern[6]
21. K-13/5	GX-5686-A	B3	+0.5	260±145	1690±145	268±145	1682±145
22. K-13/5	GX-5686-G	B3	-8.8	165±125	1785±125	170±125	1780±125
23. O-4/13	GX-5098-A	B3	+0.7	Modern[6]	Modern[6]	Modern[6]	Modern[6]
24. O-4/13	GX-5098-G	B3	-8.2	105±100	1845±100	108±100	1842±100
25. P-2/27	GX-5097-G	B3	-8.1	390±130	1560±130	402±130	1548±130
26. P-3/55	GX-5099-G	B3	-8.7	490±110	1460±110	505±110	1445±110
27. 499S466W	Beta-1925	A	-24.2	455±100	1495±100	469±100	1481±100

[1] GX (Geochron Laboratories, Cambridge, Mass.); Beta (Beta Analytic, Inc., Coral Gables, Fla.).

[2] Old (Libby) half-life (5,570 years); age referenced to year A.D. 1950; ¹³C fractionation-corrected.

[3] New half-life (5,730 years); age referenced to year A.D. 1950; ¹³C fractionation-corrected.

[4] A (bone apatite), G (bone gelatin).

[5] ND (no determination); specimen too small for analysis.

[6] Age, after ¹³C fractionation correction, completely modern.

A.D. 1450-1500 (500-450 B.P.). This age would place the site within the late prehistoric period, before the horse and gun were introduced into the region (Secoy 1953).

3. Nature and Condition of the Bison Remains

A total of 6,937 complete and fragmentary bones of modern bison were recovered from the Garnsey Site (see table 7). Of these, 2,549 (36.7%) were identifiable to specific skeletal part, 1,092 (15.7%) were indistinguishable fragments of ribs or vertebral processes, and 3,296 (47.5%) were small, unidentifiable fragments. A very few nonbison remains were also recovered, consisting primarily of butchered canids. A breakdown of the bison materials by stratigraphic level is presented in table 8. Since almost 90% of the bison sample from Garnsey derives from a series of quasi-contemporary kill events in a single archeological level (level B3), much of the subsequent discussion will treat the total bone sample as a unit.

MINIMUM NUMBER OF INDIVIDUALS

The minimum number of individuals (MNI) has been determined for the site as a whole by dividing the number of each

Table 7

Total inventory of bison remains (complete and fragmentary) from Garnsey Site

Element	N	%		Element	N	%
1. Skull	71	1.0		25. Radial carpal	19	0.3
2. Mandible	27	0.4		26. Intermediate carpal	25	0.4
3. Hyoid	13	0.2		27. Ulnar carpal	19	0.3
4. Misc. loose teeth (incisor/canine)	29	0.4		28. Accessory carpal	16	0.2
5. Misc. loose teeth (molar/premolar)	34	0.5		29. Fused 2d-3d carpal	16	0.2
				30. 4th carpal	16	0.2
6. Misc. tooth fragments	180	2.6		31. 5th metacarpal	9	0.1
7. Atlas	29	0.4		32. Metacarpal	36	0.5
8. Axis	21	0.3		33. Pelvis	92	1.3
9. Cervical (3-7)	97	1.4		34. Femur	55	0.8
10. Thoracic (1-14)	199	2.9		35. Patella	19	0.3
11. Lumbar (1-5)	64	0.9		36. Tibia	44	0.6
12. Sacrum	13	0.2		37. Astragalus	20	0.3
13. Caudal	57	0.8		38. Lateral malleolus	16	0.2
14. Unident. vertebral body	26	0.4		39. Calcaneus	24	0.3
15. Unident. vertebral pad	132	1.9		40. Naviculocuboid	20	0.3
16. Unfused vertebral summit	28	0.4		41. 1st tarsal	3	0+
17. Rib	308	4.4		42. Fused 2d-3d tarsal	13	0.2
18. Misc. rib-vertebral process fragments	1,092	15.7		43. 2d metatarsal	2	0+
				44. Metatarsal	28	0.4
19. Costal cartilage	107	1.5		45. Metapodial	9	0.1
20. Sternebra	41	0.6		46. 1st phalanx	107	1.5
21. Scapula	72	1.0		47. 2d phalanx	85	1.2
22. Humerus	44	0.6		48. 3d phalanx	74	1.1
23. Radius	32	0.5		49. Proximal sesamoid	96	1.4
24. Ulna	28	0.4		50. Distal sesamoid	34	0.5
				51. Unident. bone fragments	3,296	47.5

Note: This table should not be used to estimate minimum number of individuals, since many of the fragments listed for a particular skeletal element may derive from the same bone.

Table 8

Inventory of bison remains (complete and fragmentary) from Garnsey Site by stratigraphic level

Element	A1[1] N	A1[1] %	A2 N	A2 %	B3 N	B3 %	B4 N	B4 %	B5[2] N	B5[2] %	B6[2] N	B6[2] %
1. Skull	0	0.0	1	0.3	51	0.8	3	1.1	0	0.0	0	0.0
2. Mandible	0	0.0	2	0.6	14	0.2	4	1.5	0	0.0	0	0.0
3. Hyoid	0	0.0	0	0.0	12	0.2	0	0.0	0	0.0	0	0.0
4. Incisor/canine	0	0.0	0	0.0	21	0.3	4	1.5	0	0.0	0	0.0
5. Molar/premolar	0	0.0	1	0.3	30	0.5	9	3.3	0	0.0	0	0.0
6. Misc. tooth fragments	0	0.0	0	0.0	168	2.7	9	3.3	0	0.0	0	0.0
7. Atlas	0	0.0	2	0.6	23	0.4	2	0.7	0	0.0	0	0.0
8. Axis	0	0.0	1	0.3	19	0.3	1	0.4	0	0.0	0	0.0
9. Cervical (3-7)	0	0.0	3	0.9	92	1.5	0	0.0	0	0.0	0	0.0
10. Thoracic (1-14)	0	0.0	13	4.0	177	2.9	2	0.7	0	0.0	0	0.0
11. Lumbar (1-5)	0	0.0	6	1.8	61	1.0	2	0.7	0	0.0	1	50.0
12. Sacrum	0	0.0	2	0.6	10	0.2	0	0.0	0	0.0	0	0.0
13. Caudal	0	0.0	6	1.8	48	0.8	2	0.7	0	0.0	0	0.0
14. Unident. vertebral body	0	0.0	4	1.2	20	0.3	0	0.0	0	0.0	0	0.0
15. Unident. vertebral pad	0	0.0	5	1.5	123	2.0	3	1.1	0	0.0	0	0.0
16. Unfused vertebral summit	0	0.0	0	0.0	28	0.5	0	0.0	0	0.0	0	0.0
17. Rib	0	0.0	16	4.9	285	4.6	2	0.7	0	0.0	0	0.0
18. Misc. rib-vertebral process fragments	0	0.0	54	16.4	975	15.8	34	12.5	3	60.0	0	0.0
19. Costal cartilage	0	0.0	3	0.9	103	1.7	0	0.0	0	0.0	0	0.0
20. Sternebra	0	0.0	1	0.3	39	0.6	1	0.4	0	0.0	0	0.0
21. Scapula	0	0.0	2	0.6	67	1.1	2	0.7	0	0.0	0	0.0
22. Humerus	0	0.0	2	0.6	38	0.6	1	0.4	0	0.0	0	0.0
23. Radius	0	0.0	2	0.6	27	0.4	0	0.0	0	0.0	0	0.0
24. Ulna	0	0.0	1	0.3	20	0.3	0	0.0	0	0.0	0	0.0
25. Radial carpal	0	0.0	2	0.6	18	0.3	1	0.4	0	0.0	0	0.0
26. Intermediate carpal	0	0.0	2	0.6	23	0.4	1	0.4	0	0.0	0	0.0
27. Ulnar carpal	0	0.0	2	0.6	15	0.2	1	0.4	0	0.0	0	0.0
28. Accessory carpal	0	0.0	1	0.3	15	0.2	0	0.0	0	0.0	0	0.0
29. Fused 2d-3d carpal	0	0.0	2	0.6	14	0.2	0	0.0	0	0.0	0	0.0
30. 4th carpal	0	0.0	1	0.3	13	0.2	0	0.0	0	0.0	0	0.0
31. 5th metacarpal	0	0.0	0	0.0	9	0.1	0	0.0	0	0.0	0	0.0
32. Metacarpal	0	0.0	1	0.3	31	0.5	2	0.7	1	20.0	0	0.0
33. Pelvis	0	0.0	2	0.6	84	1.4	1	0.4	0	0.0	0	0.0
34. Femur	0	0.0	2	0.6	49	0.8	0	0.0	0	0.0	0	0.0
35. Patella	0	0.0	0	0.0	16	0.3	1	0.4	0	0.0	0	0.0
36. Tibia	0	0.0	1	0.3	35	0.6	1	0.4	0	0.0	0	0.0
37. Astragalus	0	0.0	1	0.3	19	0.3	0	0.0	0	0.0	0	0.0
38. Lateral malleolus	0	0.0	0	0.0	16	0.3	0	0.0	0	0.0	0	0.0
39. Calcaneus	0	0.0	1	0.3	22	0.4	0	0.0	0	0.0	0	0.0
40. Naviculocuboid	0	0.0	1	0.3	16	0.3	0	0.0	0	0.0	0	0.0
41. 1st tarsal	0	0.0	0	0.0	3	0+	0	0.0	0	0.0	0	0.0
42. Fused 2d-3d tarsal	0	0.0	0	0.0	13	0.2	0	0.0	0	0.0	0	0.0
43. 2nd metatarsal	0	0.0	0	0.0	2	0+	0	0.0	0	0.0	0	0.0
44. Metatarsal	0	0.0	0	0.0	25	0.4	1	0.4	0	0.0	0	0.0
45. Metapodial	0	0.0	1	0.3	7	0.1	0	0.0	0	0.0	0	0.0

Table 8 (cont.)

Element	A1[1]		A2		B3		B4		B5[2]		B6[2]	
	N	%	N	%	N	%	N	%	N	%	N	%
46. 1st phalanx	0	0.0	7	2.1	93	1.5	1	0.4	0	0.0	1	50.0
47. 2d phalanx	0	0.0	6	1.8	75	1.2	0	0.0	1	20.0	0	0.0
48. 3d phalanx	0	0.0	5	1.5	68	1.1	0	0.0	0	0.0	0	0.0
49. Proximal sesamoid	0	0.0	2	0.6	92	1.5	1	0.4	0	0.0	0	0.0
50. Distal sesamoid	0	0.0	0	0.0	34	0.6	0	0.0	0	0.0	0	0.0
51. Unident. bone fragments	7	100.0	165	50.2	2,905	47.1	192	70.6	0	0.0	0	0.0
Total	7		329		6,163		272		5		2	

Note: This table should not be used to estimate minimum number of individuals, because many fragments listed for a particular skeletal element may derive from same bone. Material listed in this table derives exclusively from major excavation units (trenches 77-1, 77-2, 78-1, 78-5, 78-6 and, 78-7); items salvaged from walls of arroyo have not been included.

[1] Unexcavated bone clusters, assignable to level A1, are exposed in arroyo walls upstream from areas investigated in 1977 and 1978.

[2] Sample limited to specimens salvaged from arroyo walls; major excavation units terminated at level B4.

specific element recovered archeologically by the number of that element in the animal. The results are presented in table 9. The best estimate for the minimum number of individuals recovered in the two seasons of excavation is thirty-five, based on skulls. Higher estimates would be obtained if: (1) each stratigraphic level were treated as a discrete unit; (2) each bone cluster within a level were treated as a discrete unit; (3) the ages of cranial and postcranial elements were considered; (4) idiosyncratic metric and nonmetric characteristics of postcranial elements were considered.

TAPHONOMIC CONSIDERATIONS

Before turning to a more detailed discussion of the Garnsey bison material, a few comments are in order concerning the preservation of the bones and the degree to which they have been disturbed by fluvial action subsequent to procurement activities at the site.

Preservation

Preservation of the Garnsey remains, for the most part, is excellent (see Speth and Parry 1980:45). With the exception of horn cores, most bones, if allowed to dry out slowly, required relatively little or no subsequent treatment with preservative before removal. Costal cartilage was well preserved. In addition, delicate plaques of cartilage remained attached to the proximal ends of several scapulae.

Table 9

Minimum number of individuals (MNI) from Garnsey Site

Element	Side	N	MNI
1. Skull	-	35 (5)¹	35 (5)¹
2. Mandible	L	9 (2)	9 (2)
	R	13 (2)	13 (2)
3. Hyoid	L	5 (1)	5 (1)
	R	6	6
4. Atlas	-	29 (4)	29 (4)
5. Axis	-	21 (1)	21 (1)
6. Cervical (3-7)	-	86 (9)	18 (2)
7. Thoracic (1-14)	-	152 (5)	11 (1)
8. Lumbar (1-5)	-	45	9
9. Sacrum	-	13	13
10. Caudal	-	55	6
11. Rib	L	86 (14)	7 (1)
	R	97 (20)	7 (2)
12. Costal cartilage	-	107	4
13. Sternebra	-	41	6
14. Scapula²	L	12	12
	R	10	10
15. Proximal humerus	L	14 (2)	14 (2)
	R	18 (5)	18 (5)
16. Distal humerus	L	14	14
	R	15 (2)	15 (2)
17. Proximal radius	L	10 (1)	10 (1)
	R	13 (6)	13 (6)
18. Distal radius	L	10 (4)	10 (4)
	R	14 (5)	14 (5)
19. Ulna	L	10 (2)	10 (2)
	R	13 (3)	13 (3)
20. Radial carpal	L	6	6
	R	13	13
21. Intermediate carpal	L	14	14
	R	11	11
22. Ulnar carpal	L	11 (1)	11 (1)
	R	8	8

Table 9 (cont.)

	Element	Side	N	MNI
23.	Accessory carpal	L	11	11
		R	5	5
24.	Fused 2d-3d carpal	L	8	8
		R	7	7
25.	4th carpal	L	9	9
		R	6	6
26.	5th metacarpal	-	9	9
27.	Proximal metacarpal	L	18 (4)	18 (4)
		R	13 (4)	13 (4)
28.	Distal metacarpal	L	12 (3)	12 (3)
		R	13 (4)	13 (4)
29.	Pelvis³	L	17 (4)	17 (4)
		R	31 (8)	31 (8)
30.	Proximal femur	L	9 (2)	9 (2)
		R	8 (1)	8 (1)
31.	Distal femur	L	14 (6)	14 (6)
		R	13 (3)	13 (3)
32.	Patella	L	9	9
		R	10	10
33.	Proximal tibia	L	12 (2)	12 (2)
		R	18 (9)	18 (9)
34.	Distal tibia	L	9 (1)	9 (1)
		R	13 (2)	13 (2)
35.	Astragalus	L	10	10
		R	10	10
36.	Lateral malleolus	L	10	10
		R	6	6
37.	Calcaneus	L	10 (2)	10 (2)
		R	12 (3)	12 (3)
38.	Naviculocuboid	L	9	9
		R	11	11
39.	1st tarsal	-	3	3
40.	Fused 2d-3d tarsal	L	8	8
		R	5	5

Table 9 (cont.)

Element	Side	N	MNI
41. 2nd metatarsal	–	2	1
42. Proximal metatarsal	L	14 (3)	14 (3)
	R	10 (1)	10 (1)
43. Distal metatarsal	L	13 (1)	13 (1)
	R	11 (1)	11 (1)
44. 1st phalanx	"L"[4]	48 (3)	12 (1)
	"R"	52 (2)	13 (1)
45. 2d phalanx	"L"	47 (1)	12 (1)
	"R"	37 (3)	10 (1)
46. 3d phalanx	"L"	36 (2)	9 (1)
	"R"	37 (2)	10 (1)
47. Proximal sesamoid	"L, "C[4]	20 (1)	5 (1)
	"L, "P	29 (1)	8 (1)
	"R, "C	24	6
	"R, "P	23	6
48. Distal sesamoid	"L"	20	5
	"R"	14	4

Note: MNI (minimum number of individuals); bison only; determined by dividing number of each specific element recovered archeologically by number of that element in animal; complete limb bones are counted twice in tabulations (proximal once; distal once).

[1] Numbers in parentheses refer to number of immature elements and minimum number of immature individuals.

[2] Includes only specimens in which part of glenoid cavity and neck are preserved.

[3] Includes only specimens in which part of acetabulum is preserved.

[4] "L," "R" (side of foot; side of animal unknown); C, P (center or periphery of foot; side of animal unknown).

Even fragile horn sheaths were recovered, some still in
place over their bony horn cores, others scattered loose
among the butchering debris. Differential loss of softer,
fragile, or less dense elements owing to factors of preservation therefore does not appear to have played a major role
in site formation at Garnsey.

This conclusion may be illustrated more explicitly by
comparing the frequency of various skeletal elements recovered at Garnsey (%MNI) with the proportion that would be expected to survive on the basis of bone density (fig. 21; the
expected survival percentages are based on caribou bone densities as presented in Binford and Bertram 1977:138 and Binford 1981:217). The correlation is low (r = 0.49; N = 25),
and with the elimination of a few obvious outliers (e.g.,
skull, atlas, pelvis, and perhaps mandible), the remaining
elements display essentially no correlation.

Destruction of elements by carnivores and rodents also
appears to have been insignificant at Garnsey. Gnaw marks,
punctures, crenulated edges, and other obvious signs of
predator-scavenger damage are extremely rare on the generally well-preserved bones.

Binford (1981) has recently proposed a series of more
quantitative approaches for assessing the degree of destruction of faunal assemblages by predator-scavengers. One such
approach focuses on the degree of attrition of long bones as
reflected by the frequency of bone cylinders per MNI and

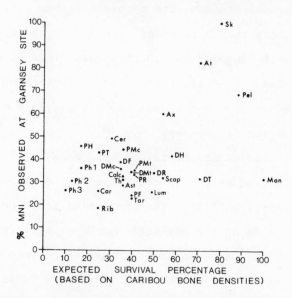

Fig. 21. Relation between proportional frequency of elements observed at Garnsey Site (%MNI) and expected survival per- centage of elements derived from caribou bone densities (from Binford and Bertram 1977:138; Binford 1981:217).

bone splinters per MNI on the site. Cylinders are major portions or segments of long-bone shafts minus the articular ends (Binford 1981:171); splinters are small shaft fragments (Binford 1978:155, 463). Gnawing animals first attack the articular ends of long bones, generating both cylinders and splinters. For a given number of bones, continued attrition then reduces the number of cylinders by gradually converting them into splinters (Binford 1981:51). The result is a graph of the sort shown in figure 22, which displays data from a series of control assemblages with known degrees of predator-scavenger destruction (Binford 1981:176). Assemblages close to the origin of the graph have suffered very little destruction; Garnsey falls clearly within this group.

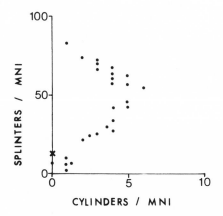

• KNOWN   CASES   OF   RAVAGED   ASSEMBLAGES
✗ GARNSEY   SITE

Fig. 22. Relation between number of bone splinters per MNI and number of bone cylinders per MNI in Garnsey assemblage compared with values in a series of control assemblages with known degrees of destruction by predator-scavengers (adapted from Binford 1981:176).

Another similar way of assessing the degree of attrition due to predator-scavengers is to compare the proportion of articular ends still part of complete bones with the frequency of cylinders per MNI (Binford 1981:176). This relation, shown in figure 23, assumes that, as destruction by gnawing animals increases, the proportion of limb bones with intact articulations decreases and the number of cylinders

increases. Control assemblages known to have suffered relatively little destruction by predator-scavengers fall at the
upper left edge of the graph; again Garnsey is clearly within this group.

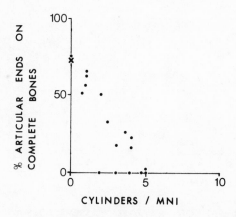

CYLINDERS / MNI

• KNOWN   CASES   OF   RAVAGED   ASSEMBLAGES
× GARNSEY   SITE

Fig. 23. Relation between frequency of articular ends remaining on complete bones and number of bone cylinders per
MNI in Garnsey assemblage compared with values in a series
of control assemblages with known degrees of destruction
by predator-scavengers (adapted from Binford 1981:176).

Both of the figures discussed above are based on the
frequency of cylinders, which are extremely rare at Garnsey.
Similar conclusions are reached, however, concerning the
limited degree of attrition of the Garnsey assemblage when
other data are considered. The likelihood that a given bone
will survive the attacks of predator-scavengers is closely

correlated with its density (Brain 1967, 1980, 1981; Binford
and Bertram 1977:138). Differences in the probability of
survival are particularly marked between the proximal and
distal ends of the humerus and tibia (Binford 1981:217-19).
The relative proportions of proximal and distal ends of
these bones on a site, therefore, provide a useful index of
the degree of attrition suffered by the assemblage. Binford
found the humerus to be particularly sensitive. Plotting a
series of control assemblages with known degrees of attrition, Binford (1981:219) obtained graphs of the type shown
in figures 24 and 25. In both figures Garnsey falls square-
ly within the "zone of no destruction."

Finally, the destruction by predator-scavengers of
bones of immature animals should be more extensive than that
of the bones of mature animals (Brain 1967, 1980; Binford
and Bertram 1977). This is clearly not the case at Garnsey.
The proportion of complete immature limb elements is substantially greater than the proportion of intact mature limb
bones (see fig. 30 below).

Fluvial Disturbance

A somewhat more complex taphonomic problem concerns the
extent to which bones have been reworked and redeposited by
fluvial action. Their association with alluvial sediments
along the flanks and bottoms of channels suggests that reworking could have played an important role not only in dis-

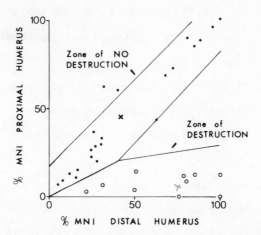

oKNOWN CASES OF RAVAGED ASSEMBLAGES
•KNOWN CASES WITH NO DESTRUCTION
×GARNSEY SITE

Fig. 24. Relation between proportional frequency of proximal humerus and distal humerus in Garnsey assemblage
(%MNI) compared with values in a series of control assemblages with known degrees of destruction by predator-
scavengers (adapted from Binford 1981:219). Note that
%MNI has been calculated in this figure according to
procedures outlined in Binford 1978:69-71.

placing materials downstream from the original kill-

processing loci, but also in sorting out and eliminating

smaller or less dense elements such as hyoids, sternebrae,

costal cartilage, rib fragments, vertebrae, and phalanges.

On the other hand, the excellent condition of the bison

skulls (many with unfused sutures), scapulae with fragile

proximal cartilage still intact, unfused epiphyses of limb

bones still in place, the absence of preferred orientations

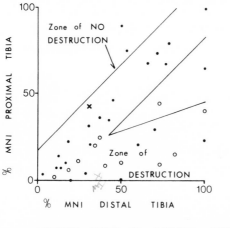

KNOWN CASES OF RAVAGED ASSEMBLAGES
KNOWN CASES WITH NO DESTRUCTION
GARNSEY SITE

Fig. 25. Relation between proportional frequency of proximal tibia and distal tibia in Garnsey assemblage (%MNI)
compared with values in a series of control assemblages
with known degrees of destruction by predator-scavengers
(adapted from Binford 1981:219). Note that %MNI has
been calculated in this figure according to procedures
outlined in Binford 1978:69-71.

of bones (strike and dip were recorded), and the fresh con-

dition of most of the bone (e.g., absence of abrasion, pol-

ish, surface exfoliation, or gnaw marks) argue for rapid

burial with little or no subsequent reworking.

One method taphonomists use to identify the nature and

extent of fluvial disturbance is to compare the observed

frequencies of skeletal elements with their expected fre-

quencies based on their dispersal potentials in flowing

water. Flume experiments have shown that the likelihood

that a bone will be transported in a stream depends largely on its density, size, and shape as well as on the velocity of the current and various channel characteristics (Voorhies 1969; Behrensmeyer 1975; Shipman 1981). Voorhies (1969) showed that bones of medium-sized animals (sheep and coyote) separated into three principal dispersal groups. The most readily transported group (group I) contained ribs, ver-tebrae, the sacrum, and the sternum. The elements least susceptible to transport (group III) were the skull and the mandible. The various limb elements fell within an interme-diate dispersal group (group II). Similar studies by Beh-rensmeyer (1975) confirmed the utility of Voorhies dispersal groups but showed that their composition varied somewhat depending on the size of the animal.

In theory, underrepresentation at Garnsey of group I and II elements, or overrepresentation of group III ele-ments, would indicate that significant fluvial disturbance and transport had occurred. Unfortunately, many of the ele-ments in groups I and II (e.g., sternum, ribs) also happen to be among the elements of highest overall food utility (see discussion of utility below) and therefore should be underrepresented at a kill site, regardless of fluvial con-ditions. Other members of group I (e.g., sesamoids) are commonly removed from kill sites as riders attached to higher-utility parts and again will be underrepresented. Similarly, group III elements (e.g., skull) are bulky, low-utility parts that typically are discarded at kills. Thus, although Voorhies dispersal groups appear to provide an objective measure of fluvial disturbance, in a kill site such as Garnsey they are ambiguous.

To better understand the extent to which the Garnsey bones have been disturbed and reworked, the configuration and content of the various bone clusters on the site were examined and compared (see Speth and Parry 1980:33-34). These investigations indicate that clusters on the south side of the original main channel are essentially intact and undisturbed. Those on the north side have been spread out somewhat in a downstream direction, but with little or no significant differential loss due to sorting.

A variety of factors point to the intact nature of the clusters on the south side of the main channel. The channel during the fifteenth century, as today, was deflected toward the north by bedrock outcrops as it passed through the site area (see figs. 3 and 4). In addition, the alluvial deposits in the wash dipped gently toward the north and west. Both of these circumstances combined to confine overbank flow primarily to the small peripheral channels on the north side of the main channel.

Other factors also argue for limited disturbance of the southern bone clusters: (1) the clusters have sharp outer boundaries; (2) there is no evidence for size sorting of materials within the clusters (the distribution of tiny resharpening flakes precisely matches that of the bones); (3) the clusters occur in silt and fine sand, not in gravel as on the north side; (4) skulls generally are intact; (5) bones are stacked directly on top of each other with no intervening lenses of alluvium; (6) unfused epiphyses are in place; (7) horn sheaths are preserved and in several cases are still on their horn cores; (8) cartilage is intact on scapulae; (9) bones are fresh and unabraded; (10) several fragments of butchered scapula blade are still stacked on top of the scapula from which they were removed; and finally (11) the southern clusters are virtually indistinguishable from each other in terms of the proportions of various skeletal elements, including smaller and less dense items (table 10).

Bone clusters on the north side of the main channel have been subjected to more fluvial disturbance than those on the south side: (1) bones are found in gravel lenses in and adjacent to shallow, braided channels; (2) cluster boundaries are not sharp, particularly on the downstream side; and (3) bones on the downstream side of clusters are often separated vertically by thin lenses of alluvium. However, for many of the same reasons given in the discussion of the southern clusters, disturbance of the northern clusters appears to have been comparatively minor and must have occurred very soon after the procurement activities at the site (e.g., fresh condition of the bones; lack of abrasion

Table 10

Inventory of identifiable skeletal elements in level B3 by excavation unit at Garnsey Site

Element	Trench 78-7 N	Trench 78-7 %	Trench 78-6 N	Trench 78-6 %	Trench 77-2 N	Trench 77-2 %	Trenches 77-2, 78-6, 78-7 N	Trenches 77-2, 78-6, 78-7 %	Trenches 77-1, 78-1, 78-5 N	Trenches 77-1, 78-1, 78-5 %
1. Skull	23	5.2	4	1.9	7	2.8	34	3.7	17	1.9
2. Mandible	2	0.4	0	0.0	2	0.8	4	0.4	8	0.9
3. Hyoid	0	0.0	2	0.9	0	0.0	2	0.2	10	1.1
4. Atlas	10	2.2	2	0.9	8	3.1	20	2.2	3	0.3
5. Axis	8	1.8	2	0.9	5	2.0	15	1.6	4	0.4
6. Cervical (3-7)	30	6.7	11	5.2	15	5.9	56	6.1	36	4.0
7. Thoracic (1-14)	53	11.9	19	9.0	36	14.2	108	11.8	65	7.3
8. Lumbar (1-5)	14	3.1	4	1.9	9	3.5	27	3.0	33	3.7
9. Sacrum	3	0.7	4	1.9	0	0.0	7	0.8	3	0.3
10. Caudal	4	0.9	6	2.8	4	1.6	14	1.5	29	3.3
11. Rib	68	15.2	40	18.9	16	6.3	124	13.6	135	15.2
12. Costal cartilage	22	4.9	9	4.2	9	3.5	40	4.4	61	6.8
13. Sternebra	9	2.0	6	2.8	4	1.6	19	2.1	19	2.1
14. Scapula	11	2.5	5	2.4	10	3.9	26	2.9	40	4.5
15. Humerus	8	1.8	9	4.2	8	3.1	25	2.7	12	1.3
16. Radius	6	1.3	3	1.4	5	2.0	14	1.5	12	1.3
17. Ulna	4	0.9	1	0.5	2	0.8	7	0.8	13	1.5
18. Radial carpal	3	0.7	5	2.4	3	1.2	11	1.2	7	0.8
19. Intermediate carpal	2	0.4	3	1.4	2	0.8	7	0.8	14	1.6
20. Ulnar carpal	4	0.9	1	0.5	3	1.2	8	0.9	7	0.8
21. Accessory carpal	2	0.4	1	0.5	2	0.8	5	0.5	10	1.1
22. Fused 2d-3d carpal	3	0.7	2	0.9	4	1.6	9	1.0	4	0.4
23. 4th carpal	3	0.7	0	0.0	4	1.6	7	0.8	6	0.7
24. 5th metacarpal	0	0.0	0	0.0	5	2.0	5	0.5	3	0.3
25. Metacarpal	5	1.1	2	0.9	4	1.6	11	1.2	15	1.7
26. Pelvis	24	5.4	16	7.5	10	3.9	50	5.5	30	3.4
27. Femur	19	4.3	5	2.4	6	2.4	30	3.3	17	1.9
28. Patella	6	1.3	2	0.9	2	0.8	10	1.1	6	0.7
29. Tibia	8	1.8	6	2.8	4	1.6	18	2.0	16	1.8
30. Astragalus	6	1.3	3	1.4	2	0.8	11	1.2	6	0.7

Table 10 (cont.)

Element	Trench 78-7		Trench 78-6		Trench 77-2		Trenches 77-2, 78-6, 78-7		Trenches 77-1, 78-1, 78-5	
	N	%	N	%	N	%	N	%	N	%
31. Lateral malleolus	5	1.1	2	0.9	0	0.0	7	0.8	9	1.0
32. Calcaneus	2	0.4	4	1.9	3	1.2	9	1.0	12	1.3
33. Naviculocuboid	5	1.1	1	0.5	0	0.0	6	0.7	10	1.1
34. 1st tarsal	1	0.2	0	0.0	0	0.0	1	0.1	2	0.2
35. Fused 2d-3d tarsal	4	0.9	2	0.9	0	0.0	6	0.7	7	0.8
36. 2d metatarsal	1	0.2	0	0.0	0	0.0	1	0.1	1	0.1
37. Metatarsal	5	1.1	4	1.9	2	0.8	11	1.2	11	1.2
38. 1st phalanx	20	4.5	8	3.8	13	5.1	41	4.5	45	5.1
39. 2d phalanx	13	2.9	7	3.3	9	3.5	29	3.2	42	4.7
40. 3d phalanx	9	2.0	8	3.8	16	6.3	33	3.6	34	3.8
41. Proximal sesamoid	17	3.8	2	0.9	17	6.7	36	3.9	52	5.8
42. Distal sesamoid	4	0.9	1	0.5	3	1.2	8	0.9	25	2.8
Total	446		212		254		912		891	

Note: This table omits loose teeth and tooth fragments, unidentifiable vertebral fragments, miscellaneous rib-vertebral process fragments, metapodials, and unidentifiable bone fragments.

and gnawing; preservation of skulls, horn sheaths, and scapular cartilage; absence, for the most part, of preferred orientations of bones; close match between distribution of bones and resharpening flakes; etc.). Finally, the frequencies of skeletal elements in the northern clusters are virtually identical to the frequencies in the undisturbed southern clusters (see table 10).

In light of these observations, it seems reasonable to proceed on the assumption that differential loss of smaller or less dense elements, due either to factors of preservation or to fluvial sorting, has not played a significant role in the formation of the archeological record at Garnsey.

4. Selective Procurement at Garnsey

SEASONALITY, AGE STRUCTURE,
AND THE SELECTIVE PROCUREMENT OF MALES

Michael Wilson (1980:93, 123) undertook a detailed analysis of the Garnsey dentitions to determine the age structure of the kill population and the season when animals were hunted within the Garnsey Wash. Combining eruption and wear data with measurements of crown heights (metaconid and paracone) of mandibular and maxillary molars, respectively, Wilson concluded that the bison were killed during the spring calving season (probably April-May, based on the timing of calving in Northern Plains bison). A somewhat earlier time for kill activities at Garnsey (late March-early April) is suggested by calving data from Southern Plains herds (Halloran and Glass 1959:369).

The age structure of the kill population indicated by the Garnsey dentitions is markedly bimodal (fig. 26) and is not typical of the expected age distribution for a catastrophic kill. Two distinct peaks are present, one at three

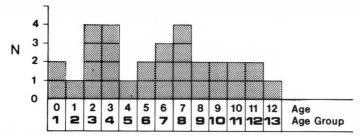

Fig. 26. Age-group distribution of bison at Garnsey Site based on combined information from mandibles and maxillae (minimum number of individuals indicated on vertical axis).

to four years and a second at seven years. Young adult animals five to six years of age are conspicuously underrepresented. Although newborn animals are frequently underrepresented at kill sites (cf. Reher 1974:121-22), either because they are more susceptible to decay or perhaps because they have been selectively removed by the hunters, the absence of five- to six-year-old adults is not so easily explained in these same terms. Instead, Wilson suggests that bimodality in the age structure is the result of superimposing two or more separate catastrophic kill events that included one or more cow-calf nursery groups and one or more bull groups. Since bull groups are composed largely of mature animals, adding one or more such groups to the age distribution of a cow-calf group might produce a bimodal age curve like the one shown in figure 26.

Wilson's argument is compatible with the archeological evidence that points to multiple kill events in the wash.

Wilson's conclusion also fits the observed sex structure of the Garnsey kill population, in which 60% of the animals are males (see below). Even during the rut the percentage of males in modern herds does not normally approach this value. And outside the rutting season the percentage of bulls usually drops considerably lower (Wilson 1980:117; McHugh 1958:14-15). Moreover, as already noted, the Garnsey dentitions argue against procurement during the rut. Thus the most plausible explanation for the high proportion of males, which also is compatible with the observed bimodal age distribution, is a pattern of multiple procurement events in which one or more cow-calf groups and several bull groups were killed in the wash. This conclusion assumes, of course, that the predominantly male sex ratio of the skulls accurately reflects the sex structure of the original kill population (all events combined), an assumption I will examine more fully below.

It should be pointed out that none of the bone clusters presumably resulting from the hunting of bull groups at Garnsey contained only bulls. As will be discussed below, most of these clusters contained a relatively high proportion of females (ca. 40%). Thus, in an average bull group of perhaps six to eight animals, two or three might be cows. The literature is somewhat unclear concerning the composition of bull groups in modern herds. McHugh (1958:14) noted that bull groups were largest during the calving season but that they rarely contained cows, other than an occasional old and barren individual. Meagher (1973:46), however, in a more recent study of the Yellowstone herd, noted that cows without calves were not uncommon in bull groups. Moreover, Halloran (1968) provides evidence from the Southern Plains that cows may not have calved every year. Thus bull groups during the calving season may often have contained a few mature females that did not give birth that year.

The proportion of noncalving females joining bull groups may in fact be an indirect reflection of the severity of local environmental conditions, since the latter would directly affect the rate of conception in the herd (Meagher 1973; Fuller 1961; Reher and Frison 1980). By this argument, the relatively high proportion of females in the Garnsey "bull groups" may indicate less-than-optimal range conditions during the mid- to late fifteenth century in southeastern New Mexico, an issue to which I will return later.

Finally, it is interesting that the procurement of bull groups containing cows that were not calving would have contributed to the underrepresentation of juvenile animals at Garnsey, despite the timing of hunts during the calving season and the presence of females in virtually every kill.

SEX STRUCTURE: DISCREPANCY BETWEEN CRANIAL AND POSTCRANIAL REMAINS

The sex of the Garnsey skulls, mandibles, and thirty-two postcranial elements has been determined using cross-

plots of various metric attributes and ratios (table 11).
Procedures used to sex the cranial materials and eleven of
the postcranial elements have been described elsewhere
(Speth and Parry 1980:81, 307-18). Additional criteria, developed specifically to help in sexing damaged or fragmentary limb elements, are presented in the Appendix. Since
the 1980 report appeared, twenty-one additional elements
have been sexed, primarily in the axial skeleton. These include cervicals 3-7; thoracics 1, 12-14; lumbars 1-5; sacrum; manubrium; metasternum (posterior, triangular sternal
body; see Kobryn 1973); pelvis; astragalus; and phalanx 2
(both front and rear). Thoracics 2-11 could not be separated with confidence and therefore were not sexed. The principal criteria used to sex these additional elements are
summarized in table 12.

Based on skulls, males outnumber females by a ratio of
60:40. Based on mandibles, females outnumber males by a
ratio of nearly 60:40. The average sex ratio for all postcranial elements combined is approximately 66:34 in favor of
females. The proportion of female bones among axial elements is lower and varies more widely than among appendicular elements. The average ratio for all axial elements combined is approximately 62:38 in favor of females, while that
for all appendicular elements combined is 70:30, also in
favor of females. In addition to skulls, the only other
elements with a sex ratio in which male parts predominate

Table 11

Sex of Garnsey cranial and postcranial material

Element	Sex	Sexed MNI N	%		Element	Sex	Sexed MNI N	%
1. Skull	M	18	60.0		21. Metasternum	M	1	25.0
	F	12	40.0			F	3	75.0
2. Mandible	M	5	41.7		22. Scapula	M	4	33.3
	F	7	58.3			F	8	66.7
3. Atlas	M	7	35.0		23. Proximal humerus	M	2	25.0
	F	13	65.0			F	6	75.0
4. Axis	M	9	50.0		24. Distal humerus	M	3	33.3
	F	9	50.0			F	6	66.7
5. Cervical 3	M	6	46.2		25. Proximal radius	M	2	33.3
	F	7	53.8			F	4	66.7
6. Cervical 4	M	5	50.0		26. Distal radius	M	3	42.9
	F	5	50.0			F	4	57.1
7. Cervical 5	M	8	47.1		27. Ulna	M	5	50.0
	F	9	52.9			F	5	50.0
8. Cervical 6	M	10	62.5		28. Proximal metacarpal	M	3	27.3
	F	6	37.5			F	8	72.7
9. Cervical 7	M	1	10.0		29. Distal metacarpal	M	3	27.3
	F	9	90.0			F	8	72.7
10. Thoracic 1	M	6	42.9		30. Pelvis (right)	M	12	60.0
	F	8	57.1			F	8	40.0
11. Thoracic 12	M	1	25.0		31. Proximal femur	M	1	16.7
	F	3	75.0			F	5	83.3
12. Thoracic 13	M	2	25.0		32. Distal femur	M	1	16.7
	F	6	75.0			F	5	83.3
13. Thoracic 14	M	2	25.0		33. Proximal tibia	M	2	28.6
	F	6	75.0			F	5	71.4
14. Lumbar 1	M	2	20.0		34. Distal tibia	M	3	37.5
	F	8	80.0			F	5	62.5
15. Lumbar 2	M	0	0.0		35. Astragalus	M	2	20.0
	F	6	100.0			F	8	80.0
16. Lumbar 3	M	2	50.0		36. Calcaneus	M	2	22.2
	F	2	50.0			F	7	77.8
17. Lumbar 4	M	4	40.0		37. Proximal metatarsal	M	3	27.3
	F	6	60.0			F	8	72.7
18. Lumbar 5	M	5	55.6		38. Distal metatarsal	M	3	27.3
	F	4	44.4			F	8	72.7
19. Sacrum	M	7	53.8		39. Phalanx 2 (front)	M	6	60.0
	F	6	46.2			F	4	40.0
20. Manubrium	M	2	33.3		40. Phalanx 2 (rear)	M	5	50.0
	F	4	66.7			F	5	50.0

Table 12

Criteria for sexing postcranial skeletal elements

Element	Criteria[1]	References
1. Cervical (3-7)	BFcr x HFcr:BFcd x HFcd BPacr:BPacd BPacr:BFcd x HFcd BPacd:BFcd x HFcd BPacr x BPacd:BFcd x HFcd	von den Driesch (1976:72-73) " " " "
2. Thoracic (1, 12-14)	BFcr x HFcr:BFcd x HFcd BPtr:BPacd BPtr:BFcd x HFcd BPtr:BPacr x BPacd	von den Driesch (1976:72-73) " " "
3. Lumbar (1-5)	BFcr x HFcr:BFcd x HFcd BPacr x BPacd:GLPa BFcd x HFcd:GLPa	von den Driesch (1976:72-73) " "
4. Sacrum	BFcr x HFcr:PL (1st body only)	von den Driesch (1976:71)

5. Pelvis	Acetabular volume (AV)
	LA:AV
	Maximum length of cranial
	  branch of pubis from
	  acetabular rim to sym-
	  physis
	Maximum thickness of pubis
	  at symphysis
	SH x SB:AV
		Kobrynczuk (1976:39)
		von den Driesch (1976:83)
		— — —
		
		— — —
		
		von den Driesch (1976:83)

6. Manubrium	IJ	Kobryn (1973:316-17, 324-27)
7. Metasternum	KL	Kobryn (1973:316-17, 324-27)
8. Astragalus	GLY x GB:Bd Volume	von den Driesch (1976:88-89) Kobrynczuk and Kobryn (1973:291)
9. Phalanx 2	Index:distal width Index:distal width	Duffield (1970:79-83, 1973) Empel and Roskosz (1963:274-75)

[1]Colon denotes crossplot; "x" denotes product; for additional criteria, see also Appendix in this study and Speth and Parry (1980:81, 307-18).

are cervical 6, lumbar 5, sacrum, pelvis, and phalanx 2 (front).

The striking disparity in the sex ratio between skulls and most other skeletal elements is not likely to be an artifact of sampling bias, nor does it reflect problems in sexing postcranial material. Smiley (1979), for example, using an independent set of measurements, arrived at nearly identical sex determinations for various forelimb elements from the Garnsey Site. If anything, for many of the elements there should be a bias in favor of males rather than females. For example, fragmentary limb elements, in which the degree of fusion of the epiphyses is unknown, can be sexed as male if the requisite metric attributes place them clearly within the male mode. Such specimens cannot be classified as female, however, if they fall within the female mode, because they may represent very immature males. Thus fragmentary female specimens are more likely to be excluded than fragmentary male specimens.

Considering the disparity in sex ratio between skulls and most postcranial elements, the question naturally arises of which element most closely approximates the sex ratio of the original kill population. An obvious source for the disparity is that female skulls may have been selectively butchered or removed from the site, thereby biasing the sex ratio of the remaining skulls in favor of males. While selective butchering of female skulls to gain access to the

brain is plausible because it is easier to open the vault of
the female, there is no convincing evidence that this was
regularly done at Garnsey. Skulls, regardless of sex, show
relatively little evidence of butchering. Furthermore,
cranial fragments are relatively scarce at Garnsey, making
it unlikely that a large number of skulls were demolished by
butchering. Of those fragments that did occur on the site,
many could still be sexed, and the majority of these were
male.

Selective removal from the site of female skulls is
somewhat more difficult to rule out. As I will show in more
detail below, there is a relatively strong negative correlation at Garnsey between the general utility of an element
(Binford 1978) and its frequency on the site expressed as a
percentage of the expected number. Thus an element of low
overall utility, such as the skull, is found in proportionately large numbers. The skull, in fact, was the most numerous element at Garnsey, providing the best estimate for
the MNI. There is also a strong tendency at Garnsey for
discrimination against female anatomical parts to increase
with the utility of the part. These two facts together suggest that the lower the general utility of an element, the
more closely its observed frequency on the site approaches
the expected number, and the more closely the sex ratio of
the element approaches the sex ratio of the original kill

population. These issues will be addressed more fully below.

In summary, I argue that, because of their low overall utility, both male and female skulls normally were abandoned at the kill site rather than being transported elsewhere for processing or consumption. I also argue that on-site processing of skulls was minimal and did not lead to significant differential destruction of female parts. These considerations lead to the conclusion that the sex ratio of the skulls represents a reasonably close approximation of the sex ratio of the original kill population, 60:40 in favor of males (all kill events combined).

METHOD OF PROCUREMENT

The following observations are relevant to understanding the method of bison procurement employed at the Garnsey Site: (1) several different kill events are represented at the site; (2) all these kills appear to have taken place in the spring; (3) these kills involved both cow groups and bull groups; (4) the bone clusters occur out on the floor of the wash along a major, perhaps perennially moist channel rather than close to the bedrock walls of the wash or within the narrower tributaries of the wash; (5) the bone clusters occur in groups, widely separated from similar groups (at least two such groups were excavated in level B3, one on the north side of the arroyo, the other about 80 to 100 m down-

stream on the south side; one or more such groups also exist
farther upstream); (6) hunting was done on foot, without the
aid of firearms; (7) projectile points are scarce (ten
points for more than thirty-five animals).

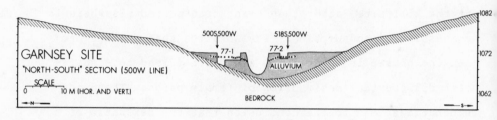

Fig. 27. North-south section of Garnsey Wash through
eastern end of trenches 77-1 and 77-2 (see fig. 7 for
approximate location), showing general configuration
of wash and position of principal bone-bearing level
(level B3; denoted by series of x's).

The methods by which the bison were gathered, contained, and killed at Garnsey remain somewhat unclear.
There is no evidence that the site was an arroyo trap. Bison remains do not occur in or near any of the major tributaries in the vicinity of the site, and it seems unlikely
that the carcasses would have been dragged great distances
from more remote trap localities. Nor was the site a classic jump locality in which the animals would have been
killed or disabled by the fall from a cliff. The walls of
the wash in the site area are neither high enough nor steep
enough, and they offer no effective obstacle to the movement
of animals in and out of the wash (see figs. 4 and 5 above).

The unsuitable nature of the wash as a jump locality is brought out more clearly in figure 27, which presents a north-south section of the site through the eastern end of trenches 77-1 and 77-2 (see fig. 7 for approximate location of section).

A variant of the classic cliff jump that could have been used at Garnsey was a stampede over the edge of the wash into some form of pound or corral (see Frison 1978: 243ff. for a description of this method). One of the earliest Spanish expeditions to explore the Pecos Valley observed a large corral or pen somewhere near Carlsbad, New Mexico, that may have been used for precisely this purpose: "The twenty-third [of November 1590] we left this place, where the river turned sharply toward the west, and we came upon a very large corral used by the Indians for enclosing cattle" (Castaño de Sosa in Hammond and Rey 1966:260-61).

If corral trapping was in fact employed at the Garnsey Site, the spatial distribution of bones in the wash indicates that several different structures would have been in operation during the period represented by levels A2, B3, and B4. Unfortunately, we encountered no direct evidence for corrals during the two seasons of excavation, but it must be noted that the site was not sampled with this specific objective in mind.

A more opportunistic method of procurement that may have been used at Garnsey, and one more compatible with the

data at hand, was cooperative ambushing or surrounding of
animals grazing along the channel or attracted to the area
by the spring a few hundred meters upstream from the site.
This method would account for the occurrence of bone clusters in widely separated parts of the wash.

Whichever method was employed--trapping animals in a
corral or surrounding them on the floor of the wash--the
scarcity of projectile points, complete or fragmentary, is
problematic. Frison (1978:243ff.) notes that large numbers
of points are usually recovered from kill sites that did not
involve a lethal jump. Projectile points at Garnsey, of
course, may simply have been salvaged from the carcasses
during the butchering. It is also possible, however, that
the points used at the kill were made of a perishable material such as wood (or even the neck tendon of bison;
cf. Weitzner 1979:240; see also comments in Medicine Crow
1978:251). Wood in particular appears to have been widely
used in the historic Southwest. Luxan's account of the Es-
pejo expedition of 1582, for example, mentions some sort of
spear or javelin that may have been made entirely of wood:
"This place we called El Arroyo de las Garrochas, because we
found many goad sticks with which the Indians kill the buffalo" (in Hammond and Rey 1966:207). Mason (1894) documents
the widespread use of arrows tipped only with wood. He
provides the following example of the existence of wooden
points among the Apache: "The arrow of the Apache sometimes

terminates in a triangular piece of hard wood, which seems
to be perfectly effective as a weapon" (letter by J. G.
Bourke, quoted in Mason 1894:669). Opler offers a similar
example of the use of wooden points among the Apache: "When
flints are to be affixed, the shaft is split, the arrowhead
is inserted, and the shaft is tightly wrapped with moistened
sinew. More commonly, however, no flint is used; the wooden
tip of the arrow is simply sharpened and fire hardened" (Opler 1965:389). Unfortunately, no direct evidence of wooden
points has been found at Garnsey. Thus there is at present
no satisfactory explanation for the unexpectedly small number of projectile points recovered from the kill.

5. Selective Processing at Garnsey

INTRODUCTION

It is common practice in kill-site reports to present a table or graph displaying the observed versus the expected frequency of the various skeletal elements recovered. These data indicate which anatomical parts were transported away from the kill, were fragmented beyond recognition by on-site processing (e.g., bone-grease rendering), or were lost through various taphonomic processes (e.g., fluvial sorting).

Observed versus expected frequencies for the Garnsey material, based on a sitewide MNI value of thirty-five, are presented in figure 28. It is evident from these data that skulls, atlases, axes, and pelvises are among the elements most commonly abandoned at the site. These are bulky or low-yield parts, and their abundance therefore is not unexpected. Ribs, costal cartilage, sternebrae, hyoids, and caudals are among the elements least well represented on the site. These elements are particularly susceptible to loss

through decay, carnivore destruction, or fluvial sorting.
However, as discussed above, taphonomic factors do not appear to have played a major role at Garnsey in altering the
proportional representation of smaller or less dense skeletal elements. Preservation of bone, including cartilage, is
excellent. Furthermore, these same elements are comparably
infrequent in clusters on the south side of the main channel, where virtually no fluvial disturbance occurred. Thus
their scarcity probably reflects in large part their selective removal by the hunters themselves. The scarcity of
caudals may, at least in part, reflect the removal of hides.
It is worth noting in this regard that caudals are better
represented in the secondary processing area than any other
identifiable vertebral element (see table 1 above). The
other low-frequency elements noted above derive from some of
the choicest meat cuts in the bison (i.e., ribs, briskets,
and tongues), and their scarcity on the site therefore is
entirely expectable (for a survey of the ethnographic and
historic literature concerning preferred meat cuts in bison,
see Wheat 1972).

Thus, at least some of the skeletal elements displayed
in figure 28 are ordered in terms of the relative value or
utility of the meat cuts from which they derive. But such
assessments of utility are too subjective and limited to be
useful in evaluating the extent to which the entire array of
elements in figure 28 may be similarly ordered. Clearly, a

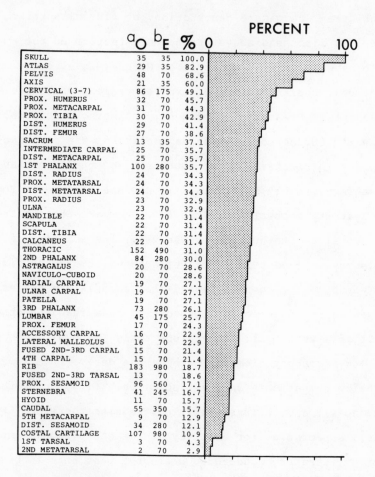

	^{a}O	^{b}E	%
SKULL	35	35	100.0
ATLAS	29	35	82.9
PELVIS	48	70	68.6
AXIS	21	35	60.0
CERVICAL (3-7)	86	175	49.1
PROX. HUMERUS	32	70	45.7
PROX. METACARPAL	31	70	44.3
PROX. TIBIA	30	70	42.9
DIST. HUMERUS	29	70	41.4
DIST. FEMUR	27	70	38.6
SACRUM	13	35	37.1
INTERMEDIATE CARPAL	25	70	35.7
DIST. METACARPAL	25	70	35.7
1ST PHALANX	100	280	35.7
DIST. RADIUS	24	70	34.3
PROX. METATARSAL	24	70	34.3
DIST. METATARSAL	24	70	34.3
PROX. RADIUS	23	70	32.9
ULNA	23	70	32.9
MANDIBLE	22	70	31.4
SCAPULA	22	70	31.4
DIST. TIBIA	22	70	31.4
CALCANEUS	22	70	31.4
THORACIC	152	490	31.0
2ND PHALANX	84	280	30.0
ASTRAGALUS	20	70	28.6
NAVICULO-CUBOID	20	70	28.6
RADIAL CARPAL	19	70	27.1
ULNAR CARPAL	19	70	27.1
PATELLA	19	70	27.1
3RD PHALANX	73	280	26.1
LUMBAR	45	175	25.7
PROX. FEMUR	17	70	24.3
ACCESSORY CARPAL	16	70	22.9
LATERAL MALLEOLUS	16	70	22.9
FUSED 2ND-3RD CARPAL	15	70	21.4
4TH CARPAL	15	70	21.4
RIB	183	980	18.7
FUSED 2ND-3RD TARSAL	13	70	18.6
PROX. SESAMOID	96	560	17.1
STERNEBRA	41	245	16.7
HYOID	11	70	15.7
CAUDAL	55	350	15.7
5TH METACARPAL	9	70	12.9
DIST. SESAMOID	34	280	12.1
COSTAL CARTILAGE	107	980	10.9
1ST TARSAL	3	70	4.3
2ND METATARSAL	2	70	2.9

^{a}O (Observed)
^{b}E (Expected)

Fig. 28. Observed versus expected number of skeletal elements at Garnsey Site (based on MNI estimate of thirty-five).

broader and more objective set of criteria is needed that can be applied to all anatomical parts of the bison.

THE MODIFIED GENERAL
UTILITY INDEX

Binford (1978), in a study of Nunamiut Eskimo hunting and processing strategies, has developed a series of explicit, quantitative utility indexes for each element in the skeletons of caribou and sheep. These indexes were derived by first determining the meat, marrow, and grease yield of each element (expressed as meat, marrow, and grease utility indexes). These values were then combined into a single general utility index (GUI) for each animal. Since certain low-utility elements may be removed from the kill as riders attached to elements of higher utility, the GUI was adjusted to produce a modified general utility index (MGUI).

The indexes developed by Binford express the utility of a given element relative to other elements in the same carcass (or similar ones). They are not intended to measure differences between carcasses arising from sexual dimorphism, nutritional state, or age. Thus, femurs from bulls and cows are assigned the same MGUI value of 100, despite the fact that the meat yield from the male femur is undoubtedly greater than that from the female. In order for identical utility values to hold for both males and females, the proportional distribution of bulk in the two sexes must be similar. Lott (1974:383) notes that bison males differ

somewhat in shape from females. In particular, bulls have
proportionately larger humps and thicker necks. Despite
these differences, the overall distribution of bulk in
adults of the two sexes appears to be broadly comparable:
the average percentage of dressed carcass weight in the
front quarters of male bison is between 53.2 and 53.9, while
that in the front quarters of females is between 52.4 and
52.5 (Peters 1958:88).

Binford (1978:475) employed variants of the caribou
MGUI to evaluate the bone frequencies from two bison kills,
Glenrock in Wyoming and Bonfire Shelter in Texas. He argued
that, in the absence of explicit utility indexes for bison,
using caribou indexes as approximations was not unreasonable, in light of the overall proportional similarity in the
anatomy of caribou, bison, and many other ungulates.

The modified general utility index derived by Binford
(1978:74, table 2.7, column 1) for caribou will be used here
to evaluate the frequency of skeletal elements recovered
from the Garnsey Site (see table 13). Used cautiously, the
MGUI should provide a reasonable approximation of the extent
to which processing decisions at Garnsey were predicated on
utility. Explanations of deviations from the ordering predicted by the MGUI must be presented tentatively, however,
since it may be difficult at times to discriminate between
deviations that stem directly from processing decisions by
prehistoric hunters and deviations due to inadequacies in

using for bison an index designed for caribou. Nevertheless, the prospect of gaining further insight into the nature and causes of the variability in bone frequencies observed at Garnsey makes the attempt worthwhile.

PROCESSING DECISIONS AT GARNSEY:
THE TREATMENT OF BOTH SEXES COMBINED

Figure 29 shows the relationship between the MGUI and
the frequency of skeletal elements at the Garnsey Site expressed as a percentage of the expected number (%MNI) for an
MNI of thirty-five animals (position and shape of "regression line" are approximate). The curve displays a relatively strong negative curvilinear relationship for most elements. The shape of the curve is of interest. The frequency of most elements is confined to a rather narrow range
(ca. 25-35%) over a wide range of utility values (from ca.
15-20 to 100). Only below utility values of about 15 does
the frequency of elements rise sharply; these low-utility
elements (skull, atlas, axis) were discarded in large numbers on the site. The shape of the distribution in figure
29 corresponds to what Binford (1978:81) has termed a "bulk"
curve, in which hunters "select for large quantities of
parts of both high and moderate value and abandon parts of
the lowest utility at rapidly accelerating rates." The opposite extreme is the "gourmet" curve in which hunters select only the highest-utility parts and discard large quantities of moderate- and low-value parts.

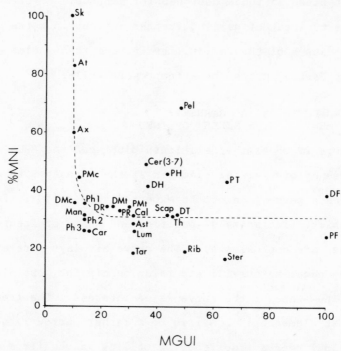

Fig. 29. Relation between proportional frequency of elements abandoned on Garnsey Site (%MNI) and modified general utility index (MGUI) (from Binford 1978:74, table 2.7, column 1).

Five skeletal elements in figure 29 deviate noticeably from the general trend of the curve; these are the pelvis, cervicals, humerus, proximal tibia, and distal femur. These elements were abandoned at the site in frequencies greater than would be predicted on the basis of their utility values alone. Pelvises, apparently because of their bulk, were stripped of meat and discarded in large numbers on the site.

The treatment of the pelvis will be considered more fully below.

Figure 29 also shows that higher-than-expected proportions of neck and upper forelimb parts were left behind at Garnsey. Similar discrimination was observed by Binford (1978:40) in the Nunamiut treatment of caribou. Binford noted "an anatomical bias against front legs and necks. These are the parts considered by the Eskimo to be most responsive to nutritional variability and are thus considered less 'reliable' parts, particularly in spring." A good case can be made for nutritional stress affecting the Garnsey bison, particularly the calving females, in the spring. This will be addressed more fully below.

The higher-than-expected proportion of proximal tibiae may also reflect discrimination by the hunters against an anatomical part of calving females rendered less desirable by nutritional stress in the spring. The tibia is relatively important as a marrow bone (Binford 1978:42-43; Stefansson 1925:233-34). The composition of the marrow in the proximal and distal ends is markedly different, however. The proximal end has a lower total fat content as well as a markedly lower content of oleic acid, the principal fatty acid in the marrow fat (Dietz 1946; Irving, Schmidt-Nielsen, and Abrahamsen 1957; Meng, West, and Irving 1969; West and Shaw 1975; Binford 1978:42; Turner 1979). The marrow of the proximal tibia apparently also is less palatable (Binford

1978:42; Stefansson 1925:233-34; Wilson 1924:174). As I
will argue more fully below, the marrow-fat content of the
proximal tibia may have been further reduced by nutritional
stress in the spring, particularly in calving females. In
the hind leg, the proximal tibia is one of the elements most
susceptible to depletion of marrow-fat reserves (Harris
1945:320; Ratcliffe 1980:337).

Finally, the higher-than-expected proportion of distal
femurs may also reflect late-spring discrimination against a
nutritionally poor female marrow bone, much as in the case
of the proximal tibia. The composition and response to
stress of marrow in these two elements are similar. Comparable discrimination against the proximal femur of females, though not evident in figure 29, was also marked at
Garnsey and will become apparent in later discussion.

The importance of general utility in processing decisions at Garnsey may be seen in differences in the degree of
breakage of limb bones from the front and hind legs. Figure
30 shows the frequency of complete, mature (fused) limb elements, expressed as a percentage of the total number of
identifiable specimens of that element, plotted against the
MGUI values. Since the utility values of the proximal and
distal portions of an element differ, the higher of the two
values was arbitrarily selected in constructing the figure.
Comparable results are obtained using the lower values. The
relations in figure 30 are clear-cut. First, within each

limb, front or rear, the proportion of complete elements abandoned on the site increases as one moves down the limb (i.e., as one moves toward lower general utility values). Second, with the exception of the metatarsal, the proportion of complete elements from the hind limbs is considerably lower on the site than the proportion of complete elements from the front limbs, again as predicted by general utility.

It is interesting that Binford's meat utility index (1978:23) and marrow index (1978:27), two of the principal indexes upon which the MGUI is based, are negatively correlated, owing particularly to the increase in oleic acid in the marrow of the lower-limb elements (r = -0.42, limb elements only; N = 13). Figure 30 therefore indirectly shows that the amount of breakage at Garnsey decreases with increasing marrow utility. This implies that much of the observed bone breakage may have been the result of butchering practices rather than on-site marrow processing, unless only the least-desirable, low-oleic-acid marrow bones were being broken open for on-site snacking. This conclusion is also supported by the fact that the marrow index is by far the poorest predictor of limb-element frequencies at Garnsey, as shown in table 14.

There is a growing debate in the literature concerning the extent to which prehistoric hunters removed anatomical parts from carcasses primarily by cutting elements apart at points of articulation or instead by chopping directly

through bones, breaking them up in the process (Binford
1981:142-47; Frison 1970; Lyman 1978; Read 1971:53, 67;
White 1952:338). While the Garnsey data are far from con-
clusive, they do suggest that bone chopping during butchering may have contributed to the pattern of breakage observed. It should be pointed out, however, that no stone
choppers or hammers were recovered, and very few of the
broken limb bones are convincing tools.

DISCRIMINATION AGAINST
FEMALES IN PROCESSING

When the sex of elements abandoned at the Garnsey Site
is taken into consideration, differential treatment of males
and females becomes readily apparent. Figure 31 displays
the relation between %MNI and the MGUI for each male and female element sexed (position and shape of "regression line"
are approximate). In this figure %MNI is calculated in-
dependently for each sex using eighteen skulls as the basis
for the MNI of males and thirteen atlases as the basis for
the MNI of females. The data used in figure 31 are summarized in table 13.

Both the male and the female curves in the figure have
similar shapes of the "bulk" variety described above. For
both sexes, large numbers of elements of high and intermediate utility have been removed from the site, and the low-
utility parts have been discarded. That elements of both
sexes tend to form patterned distributions of similar gener-

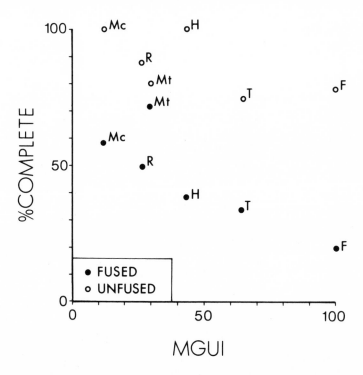

Fig. 30. Relation between proportional frequency of complete (mature and immature) limb elements (%Complete) and modified general utility index (MGUI) (from Binford 1978:74, table 2.7, column 1).

al configuration lends support to the value of Binford's caribou indexes for investigating bison. The overall similarity of the male and female curves in figure 31 also indicates that the utilities of various anatomical parts of the female were ranked in much the same manner as were the male parts, despite the high degree of sexual dimorphism in bison. This supports the contention made earlier that the

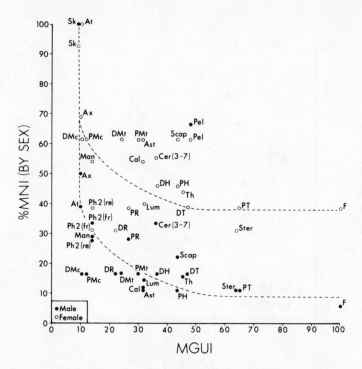

Fig. 31. Relation between %MNI (male) and %MNI (female) and modified general utility index (MGUI) (from Binford 1978:74, table 2.7, column 1). Data summarized in table 13.

proportional distribution of bulk in the fore- and hindquarters of male and female bison is broadly comparable (Peters 1958).

Examination of figure 31, however, reveals that a more complex pattern of decision making was operating at Garnsey during the butchering and processing of the kill. The female curve is considerably higher than the male curve, indicating that a greater proportion of virtually all female

Table 13

Observed proportions of male and female elements, based on MNI
estimates of eighteen bulls and thirteen cows (see table 11)

Element	%MNI (male)	%MNI (female)	%MNI (female) Minus %MNI (male)	MGUI [1]
1. Skull	100.0	92.3	-7.7	8.74
2. Mandible	28.8	53.8	+25.0	13.89
3. Atlas	38.9	100.0	+61.1	9.79
4. Axis	50.0	69.2	+19.2	9.79
5. Cervical (3-7)	33.3	55.4	+22.1	35.71
6. Thoracic (1, 12-14)	15.3	44.2	+28.9	45.53
7. Lumbar (1-5)	14.4	40.0	+25.6	32.05
8. Pelvis (right)	66.7	61.5	-5.2	47.89
9. Sternum [2]	11.1	30.8	+19.7	64.13
10. Scapula	22.2	61.5	+39.3	43.47
11. Proximal humerus	11.1	46.2	+35.1	43.47
12. Distal humerus	16.7	46.2	+29.5	36.52
13. Proximal radius-ulna [3]	27.8	38.5	+10.7	26.64
14. Distal radius-ulna [4]	16.7	30.8	+14.1	22.23
15. Proximal metacarpal	16.7	61.5	+44.8	12.18
16. Distal metacarpal	16.7	61.5	+44.8	10.50
17. Proximal femur	5.6	38.5	+32.9	100.00
18. Distal femur	5.6	38.5	+32.9	100.00
19. Proximal tibia	11.1	38.5	+27.4	64.73
20. Distal tibia	16.7	38.5	+21.8	47.09
21. Astragalus	11.1	61.5	+50.4	31.66
22. Calcaneus	11.1	53.8	+42.7	31.66
23. Proximal metatarsal	16.7	61.5	+44.8	29.93
24. Distal metatarsal	16.7	61.5	+44.8	23.93
25. Phalanx 2 (front)	33.3	30.8	-2.5	13.72
26. Phalanx 2 (rear)	27.8	38.5	+10.7	13.72

[1] MGUI (modified general utility index)(from Binford 1978:74, table 2.7, column 1).
[2] %MNI (male) and %MNI (female) values based on manubrium only.
[3] %MNI (male) and %MNI (female) values based on proximal ulna only.
[4] %MNI (male) and %MNI (female) values based on distal radius only.

parts, regardless of utility, was left behind at the site.
This implies that hunters were much more selective, across
the board, in handling female carcasses. Moreover, while
both male and female parts form distinct curvilinear distributions with respect to the MGUI, the points representing
female parts are more scattered or dispersed with respect to
their regression line than are those for males. This is
brought out more clearly by transforming (linearizing) the
distributions using common logarithms and calculating correlation coefficients (table 14). These results suggest
that the proportions of various male anatomical parts removed from Garnsey conform more closely to the expectations
of the MGUI than do the proportions of female parts taken
from the site.

One possible explanation for the greater "noise" in the
distribution of female parts is that the data reflect decisions concerning two different subsets of animals--calving
and noncalving cows. As I will discuss more fully below,
the butchering decisions at Garnsey appear to have been
closely tied to the animals' nutritional condition, which in
a spring kill would have been poorest in calving females.

Not only were more female than male parts abandoned on
the site, but fewer of the discarded female elements have
been broken in butchering and/or on-site marrow processing.
Considering the same six limb elements shown in figure 30,
only 75.0% of the sexable mature (fused) male elements are

Table 14

Correlation between utility indexes (Binford 1978) and proportional frequency of male and female skeletal elements at
Garnsey Site (frequency data transformed to common logarithms)

Utility Indexes	Log_{10}%MNI (male)	Log_{10}%MNI (female)
All elements(24)[1]		
MGUI	-0.66	-0.45
Meat	-0.50	-0.36
Grease	-0.54	-0.29
Marrow	-0.28	-0.13
Limb elements(13)[1]		
MGUI	-0.85	-0.47
Meat	-0.82	-0.37
Grease	-0.58	-0.36
Marrow	0.15	0.12

[1]Values in parentheses indicate number of elements used.

complete, whereas 98.5% of the sexable mature female elements are complete. This difference may to some extent
reflect difficulties in sexing fragmentary female specimens;
but, given the magnitude of the difference in breakage between the two sexes, these data probably do reflect a real
pattern of discrimination against females by the prehistoric
hunters.

The nature of the discrimination against female parts
can be examined more closely from another perspective. Figure 31 shows that the male and female curves are closest
together at the low end of the utility scale, diverge in the
intermediate range, then remain widely separated, but more

or less parallel, in the upper utility range. This pattern
indicates that the higher the general utility value of an
anatomical part, the greater the discrimination against the
female part.

This trend is brought out more sharply by subtracting
the %MNI of the male from that of the female for each element and plotting the difference against the MGUI (table 13;
fig. 32). Most elements appear to conform to a reasonably
tight curvilinear pattern (position and shape of "regression
line" are approximate). The curve rises rapidly at low MGUI
values, indicating that the difference between the %MNI of
females and that of males is increasing (i.e., the discrimination against the female element increases with the general
utility of the element). At intermediate and high MGUI values the curve levels off, and the discrimination against female parts, while remaining high, becomes more or less constant.

These results indicate that, when both male and female
carcasses are present (e.g., bull groups containing adult
females without calves; cow-calf groups containing younger
bulls), the hunters evaluate the utility of an anatomical
part not only with respect to other parts in carcasses of
the same sex, but also with respect to the same part in carcasses of the other sex. Age and nutritional state of the
carcasses also enter into the decision process (see discussion below). Ultimately, in an animal as sexually dimorphic

as the bison, a series of utility indexes is needed that in-corporates all these dimensions.

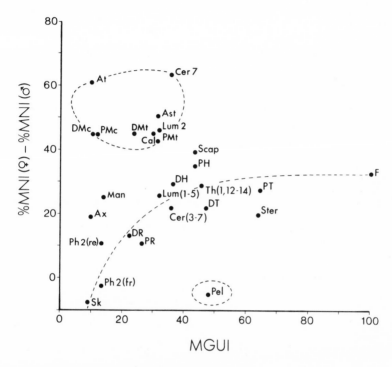

Fig. 32. Relation between %MNI (female) minus %MNI (male) and modified general utility index (MGUI) (from Binford 1978:74, table 2.7, column 1). Data summarized in table 13.

Several elements of low to moderate utility form a dis-tinct cluster above the curve in figure 32. These include the metapodials, atlas, astragalus, calcaneus, cervical 7, and lumbar 2. All these items reflect extreme discrimina-tion against the female; that is, an unexpectedly high pro-

portion of these elements left behind at the site were female. A single element, the pelvis, falls well below the
curve. In this case virtually no discrimination by sex was
made; male and female alike were dumped on the site. These
elements deserve further comment. The limb elements will be
considered first (metapodials, astragalus, and calcaneus),
followed by a discussion of the axial elements (atlas, cervical 7, lumbar 2, and pelvis).

FAT DEPLETION IN
FEMALES UNDER STRESS

I suggest here that the unexpectedly high degree of
discrimination against female metacarpals and metatarsals
reflects the poor nutritional state of females at the time
of the kill (spring calving season). The metapodials are
particularly important as marrow bones. They produce relatively little meat, but they have large marrow-cavity vol-
umes and, under normal conditions, high fat content
(cf. Binford 1978:23). Marrow from the metapodials has a
considerably higher total fat content than marrow from the
upper portions of the limbs (Dietz 1946). Moreover, the fat
in the lower part of the limbs differs in composition from
that in the upper part. Marrow fat in the metapodials and
phalanges is composed largely of oleic acid, an unsaturated
fatty acid with a low melting point (Irving, Schmidt-
Nielsen, and Abrahamsen 1957; Meng, West, and Irving 1969;
West and Shaw 1975; Binford 1978:23-25; Turner 1979). The

proportion of low-melting-point fat decreases as one moves up the limbs, reaching its lowest values in the proximal humerus and proximal femur. The sharpest decline occurs in the proximal radius and proximal tibia. Normally the fat in the metapodials and phalanges is fluid, while that in the proximal radius, proximal tibia, femur, and humerus is solid.

There is limited ethnographic documentation that marrow from the metapodials is more palatable than marrow from higher up in the limbs, owing to its higher total fat content and perhaps to its higher proportion of oleic acid (Stefansson 1925:233-34; Binford 1978:23-25; Weltfish 1965: 218-19; Wilson 1924:174).

When an animal suffers from malnutrition, fat reserves stored in various parts of the body are mobilized. In many animals there appears to be a relatively consistent sequence of fat mobilization, beginning with the back fat, followed by abdominal and kidney fat, and finally marrow fat (Harris 1945; Riney 1955; Ransom 1965; Sinclair and Duncan 1972; Pond 1978).

Within the marrow bones themselves, depletion begins at the proximal end of the limbs and moves toward the feet (Ratcliffe 1980:337; Brooks, Hanks, and Ludbrook 1977). The marrow fat from the upper front leg (humerus and radius) appears to be somewhat more susceptible to depletion than that from the upper part of the rear limb (femur and tibia). The

differences are greatest between the radius and the tibia.
On the other hand, the metatarsal fat appears to be slightly
more vulnerable to loss than the metacarpal fat (data based
on roe deer; Ratcliffe 1980).

Since the marrow fat is the last reserve to be depleted, reduced marrow-fat content and lowered proportions of
oleic acid are most likely to occur, if at all, in the
spring (see Turner 1979:602; Newlin and McCay 1948; Warren
1979; Warren and Kirkpatrick 1978; Franzmann and Arneson
1976:339). Fat reserves are replaced in more or less the
reverse order, beginning with the marrow (Riney 1955).

There are very few studies comparing patterns of variability in the nutritional condition of male and female bison at different seasons, and none to my knowledge concerning fat reserves in the bone marrow. To gain some idea of
the probable general condition of male and female bison, and
in particular their marrow, at various times of year, we
must draw on studies of other animals. Certain aspects of
bison condition are sufficiently well documented to provide
useful reference points in such comparisons: (1) male bison
are in best overall condition in the late spring-early summer and in poorest condition during and after the rut; (2)
females are in poorest condition during the calving season
and in best condition in the fall and early winter (Ewers
1958:76; Wissler 1910:41; Point 1967:120, 166; Grinnell

1972:269; Earl of Southesk 1969:80; Coues 1897:577; Roe 1972:860-61; Lott 1979:429).

These limited observations indicate that seasonal changes in bison condition parallel reasonably well the changes described in many other animals. Thus the following general pattern may be suggested for bison (Allen 1979; Anderson et al. 1972; Nordan, Cowan, and Wood 1968; Peterson 1977; Pond 1978; Riney 1955; Sinclair and Duncan 1972; Binford 1978:40). After a low point in late winter-early spring, male condition improves rapidly during late spring-early summer, as the animals build up fat reserves toward a peak at the onset of the summer rut. Male condition declines sharply during and after the rut (Lott 1979; Nordan, Cowan, and Wood 1968). Lott (1979:429), for example, notes that bison males may lose up to 10% of their normal body weight. Losses may be sufficient to cause mobilization of marrow fat (Brooks 1978). During late fall-early winter, male condition may again improve somewhat, but males often enter winter in comparatively poor condition. If there are nutritional shortages during the winter, body-fat reserves begin to be depleted, and overall condition declines through late winter into spring. With the growth of new spring forage, male condition improves rapidly, beginning a new annual cycle.

Female condition varies throughout the year in a manner broadly paralleling that of the male, but with several im-

portant differences in degree and timing. Female condition
is at its lowest during the spring calving season, somewhat
later than the low point in males. By the time females
reach their minimum, males are already moving rapidly toward
their peak in condition. After the calving season, female
condition improves gradually throughout summer and fall,
perhaps with slight losses during the rut. During this extended period of improvement, females build up reserves for
winter and the following calving season. As in males, condition in females declines during late winter. Since females often enter winter with greater reserves than males,
their condition may be somewhat better in early spring.
Later in the spring, however, pregnant or lactating females
have declined to a level of condition below that of males.

If this general pattern applies to bison, then many of
the Garnsey females, which appear to have been killed during
or close to the calving season, probably were in poorer
overall condition than the males. Reduced fat reserves in
pregnant or lactating females, combined with the smaller
bulk of cows, probably account for the general pattern of
discrimination against female anatomical parts observed at
Garnsey.

To this point I have argued that discrimination against
the Garnsey females during butchering reflected the poor nutritional condition of cows that were calving or lactating.
But I also noted earlier that females in the Southern Plains

probably did not calve every year and that some of these females may have temporarily joined bull groups. Presumably,
therefore, such females would not have been suffering significantly greater nutritional stress than bulls and thus
should not have been discriminated against during butchering
with the same intensity as would calving or lactating cows.

Is there direct evidence at Garnsey that discrimination
was, in fact, greater against females in cow-calf groups
than against cows in bull groups? To answer this question,
we must first identify which bone clusters represent cow
groups and which represent bull groups. Using the skull as
the most reliable indicator of the sex ratio of the original
kill population of each procurement event, trenches 78-6 and
78-7 (combined) and trench 77-2, on the south side of the
arroyo, probably represent two level B3 bull groups (66.7%
and 50.0% male, respectively). Bison remains on the north
side of the arroyo (trenches 77-1, 78-1, and 78-5) have been
spread out somewhat by flooding, making it more difficult on
this side to distinguish discrete kill events. Probably
only a single procurement episode is represented by the
level B3 remains, a cow-calf group (33.3% male, based on
skulls), although the possibility that more than one event
is included in the sample cannot be ruled out entirely. The
%MNI of calves, based on the proportion of immature pelvises
(unfused ilium-ischium-pubis), supports these interpreta-
tions (16.7% in the presumed bull groups in trenches 78-6

and 78-7 and in trench 77-2; 40.0% in the presumed cow-calf group[s] in trenches 77-1, 78-1, and 78-5).

Figure 33 shows the proportion of discarded mature female metapodials in the two bull groups and in the cow-calf group(s). The figure strongly supports the argument that discrimination was focused primarily against calving or lactating cows, not against females in general. Discrimination was strongest against the metatarsal, which, as noted earlier, is somewhat more vulnerable than the metacarpal to loss of fat (Ratcliffe 1980).

The unexpectedly high degree of discrimination displayed in figures 32 and 33 against two important female marrow bones at Garnsey suggests that environmental conditions during the winter and spring months immediately preceding the kills had been severe enough to cause the depletion of other major body-fat reserves in pregnant or lactating members of cow-calf groups. Discrimination against the marrow bones may also indicate that preceding summer and fall pasture conditions had been poor, preventing pregnant or lactating females from building up adequate reserves of fat to make it through the winter and spring without significant loss of marrow fat (Nordan, Cowan, and Wood 1968). I assume, on the basis of the arguments presented above, that males, which presumably had also been subject to winter and early-spring nutritional stress, had rebounded by the spring

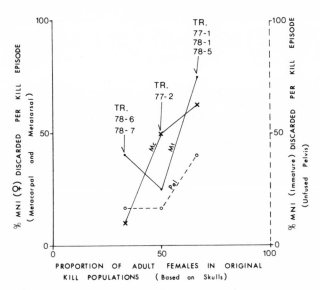

Fig. 33. Relation between proportion of adult females in
original kill population of three level B3 procurement
episodes (trenches 78-6 and 78-7; 77-2; 77-1, 78-1, and
78-5) and proportional frequency of discarded female
metapodials at Garnsey Site (%MNI female); figure also
shows relation between proportion of adult females in
original kill population of same three procurement
episodes and proportional frequency of discarded immature pelvises (%MNI; unfused ilium-ischium-pubis).

calving season to the point where their overall condition

was better than that of many of the females.

The high degree of discrimination evident in figure 32

against female astragali and calcanei (and perhaps against

other tarsals and carpals as well) probably is linked di-

rectly to the treatment of the metapodials. The discarded

metapodials were removed intact from the limbs, either by

chopping through the tibiae and radii or more commonly by
cutting through the joints; the astragali and calcanei were
removed and discarded as riders in the same process.

The unexpectedly high selectivity against the atlas,
seventh cervical, and second lumbar of female bison and the
apparent lack of discrimination against the pelvis (fig.
32) are best understood by comparing the treatment of these
elements with the handling of other parts of the axial skeleton. To accomplish this, all the vertebral elements except
thoracics 2-11 and caudals were identified and sexed. The
results are summarized in table 11 above. Criteria for
sexing the various axial elements are summarized in table
12. Thoracics 2-11 and caudals could not be distinguished
reliably and therefore were excluded from the analysis.
Figure 34b presents the results graphically, expressed as
the percentage that male parts constitute of the total number of sexed specimens for each element.

If we assume, as I argued earlier, that the sex ratio
of the original kill population is accurately reflected by
the skulls, then the distortion of this ratio at different
points along the axial skeleton owing to selective processing is clearly visible. If one ignores, for the moment, the
sharp downswings at the atlas, seventh cervical, and second
lumbar, there is a steady decrease in the proportion of male
parts left at the site as one moves from the skull toward
the rear of the ribcage. Beginning with the first lumbar

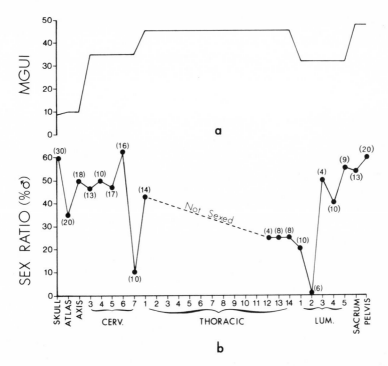

Fig. 34. MGUI values (a) and sex ratio (%male) (b) of axial elements at Garnsey Site (values in parentheses indicate number of specimens sexed).

and continuing toward the rear of the carcass, this trend is reversed.

The degree of discrimination against female axial elements covaries with MGUI values (fig. 34a). Utility values are lowest for the skull (8.74), rise to intermediate values for the cervicals (35.71), and reach a peak for the thoracics (45.53). Behind the thoracics the MGUI values decline again to a value of 32.05 for the lumbars. Despite their high utility value (47.89), the pelvises at Garnsey display no discrimination against females. Moreover, pelvises were discarded in substantial quantities at the site (nearly 70% of that expected). The pelvises were broken apart and stripped of meat on the site rather than being removed as components of larger butchering units for processing elsewhere.

The anomalous sharp downswings in figure 34b at the atlas, and especially at the seventh cervical and second lumbar, are curious, and I can offer only a tentative explanation. There is no apparent reason in terms of utility values to expect such strong selectivity by the hunters for these particular male elements. A more plausible explanation is that the downswings denote consistent points where the male vertebral column was chopped through to divide the carcass into transportable units. If so, at least one additional point probably existed within the long thoracic series. If males were preferentially butchered, and if vertebral units were consistently separated at the same points, it is plausible that many more male than female atlases, seventh cervicals, and second lumbars were destroyed. It is interesting in this regard that the downswings become more pronounced as one moves from the front of the carcass toward the rear, paralleling the trend toward increasing MGUI values. Unfortunately, it has not been possible to verify this explanation for the downswings in figure 34b. Although

there are many fragmentary vertebral bodies that clearly
have been chopped into or through by some sort of heavy cutting tool, none can be reliably identified to specific element, and none can be sexed.

Following the same logic, the upswings in figure 34b at
the sixth cervical and third lumbar may represent consistent
points in female carcasses where the spinal column was
chopped through, destroying many of the female vertebrae in
the process and elevating the proportional representation of
male elements. It is interesting that the female butchering
units so divided would have been longer than their male
counterparts, but since the female carcass is smaller, they
may have been similar to male units in total weight. These
differences in the handling of male and female vertebral
columns may reflect an attempt to maintain butchering units
of broadly comparable bulk (I am grateful to Susan L. Scott
for drawing my attention to the possible significance of the
upswings in fig. 34b).

DISCRIMINATION AGAINST IMMATURE
ANIMALS IN PROCESSING

Up to this point I have focused on selective processing
stemming from two principal sources: (1) gross differences
in the bulk of adult bulls and cows; and (2) seasonal differences in the relative nutritional state of adult bulls
and cows. Two other important sources of selectivity should
be considered: (1) differences in gross bulk and condition

between animals of the same sex at different ages; and (2) changes in the ratio of male to female bulk at different ages.

The degree of within- and between-sex discrimination at different ages is particularly difficult to investigate. First, postcranial remains can be assigned only to one of two very broad age categories (fused versus unfused or fusing). Moreover, the age of fusion differs from one element to another. Second, most unfused specimens, except for a few obvious male parts, cannot be sexed. The only direct way that was found to investigate the role of age in processing decisions at Garnsey was to examine the proportion of complete unfused bones expressed as a percentage of all complete and fragmentary unfused specimens (with or without unfused epiphyses; only proximal fragments considered). The results are displayed graphically in figure 30 above.

Figure 30 reveals two interesting aspects of the handling of younger animals. First, the proportion of complete unfused elements is considerably higher than that of complete fused elements. This suggests that younger animals were subjected to less intensive butchering and/or on-site marrow processing than adults. Second, the proportion of intact immature front-limb elements is greater than the proportion of intact rear-limb parts. This indicates that, as in adult animals, there was more discrimination against the front quarters than against the rear quarters. But the difference between front- and rear-limb values is much smaller among immature animals than among adults, suggesting that the Garnsey hunters made much less distinction between the front and rear legs of younger animals. Unfortunately, the immature elements cannot be examined by discrete age groups. It is expected that the older the animal, the greater the discrimination against front-limb elements, approaching the values seen in adults.

Two major factors probably contribute to the pattern of discrimination against younger animals during processing. The first, and most obvious, is the smaller size of immature bison. If hunters were operating under transport or time constraints, calves and yearlings may have been bypassed in favor of larger animals.

The second factor that may have favored discrimination against immature animals is their lower overall levels of subcutaneous, intermuscular, intramuscular, and marrow fat. No data are available for fat levels in bison of different ages, but several studies of cattle provide useful comparative information. For example, the percentage of total carcass fat in cattle increases markedly with age (Allen et al. 1976; Link et al. 1970a,b,c; Link et al. 1970; Hunsley et al. 1971). In addition, the proportion of fat in marrow bones from both the axial and the appendicular skeletons of cattle is substantially higher in older animals. For example, fat levels in the red or hemopoietic marrow of the

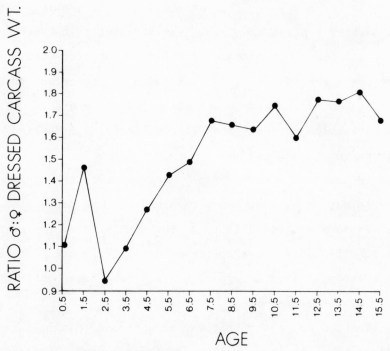

Fig. 35. Proportion of male to female dressed carcass weight in modern Southern Plains bison (data from Halloran 1961).

cervical vertebrae increased from 5.3% in calves to 14.8% in utility-grade cows to 30.8% in good-grade young bulls (Field et al. 1978; Field et al. 1980; Arasu et al. 1981). The fat content in the lumbar vertebrae also increased with age, more rapidly than in the cervicals, reaching considerably higher levels in mature animals (Field et al. 1980; Arasu et al. 1981; see also Kunsman et al. 1981). Finally, the marrow-fat content of the femur in cattle increased from between 33.7% and 47.7% in calves to levels exceeding 85% to 90% in adults (Field 1976; Kunsman et al. 1981; see also Dietz 1949; Franzmann and Arneson 1976).

Discrimination directed specifically against females can be expected to increase with increasing age. Data provided by Halloran (1961; see also Novakowski 1965) for modern Southern Plains bison indicate that the ratio of male to female dressed carcass weight increases sharply from the ages of about 2.5 to 7.5 years, after which it levels off (the adult dressing percentage in both sexes is almost identical at 54%). Halloran's data are graphed in figure 35. Unfortunately, no way was found to investigate this expectation in the Garnsey sample.

6. Southern Plains Bison in the Fifteenth Century

INTRODUCTION

The selective processing of bison at Garnsey, particularly the discarding of unusually high proportions of important marrow bones from pregnant or lactating cows, very
likely reflects the stressful conditions of spring in the
Southern Plains, when the biomass and quality of available
forage were at a minimum. The severity of the food problem
confronting bison overwintering in the region can be illustrated more clearly by examining the nutritional value of
the major forage species.

AVAILABLE FORAGE SPECIES
IN THE SOUTHERN PLAINS

Bison are grazers whose diet consists almost entirely
of grasses and some sedges (cf. Peden 1972). Forbs and
shrubs form only a minor component, generally less than 10%
by dry weight. Two of the most important forage species in
the Southern Plains are blue grama (<u>Bouteloua</u> <u>gracilis</u>) and
buffalo grass (<u>Buchloe</u> <u>dactyloides</u>). In the more xeric por-

tions of southeastern New Mexico, these grasses decline in importance, and black grama (<u>Bouteloua eriopoda</u>) and tobosa (<u>Hilaria mutica</u>) become the dominant species. Other important grasses over much of the region include side oats grama and hairy grama (<u>Bouteloua curtipendula</u> and <u>B</u>. <u>hirsuta</u>), big and little bluestem (<u>Andropogon gerardi</u> and <u>A</u>. <u>scoparius</u>), sand dropseed (<u>Sporobolus cryptandrus</u>), alkali sacaton (<u>Sporobolus airoides</u>), bush and ring muhly (<u>Muhlenbergia porteri</u> and <u>M</u>. <u>torreyi</u>), various species of three-awn (<u>Aristida</u> spp.), needle and thread (<u>Stipa comata</u>), and New Mexico feather grass (<u>Stipa neomexicana</u>)(Bureau of Land Management 1979; Castetter 1956; Gay and Dwyer 1970; Paulsen and Ares 1962; Peden 1972; Van Dyne 1973, 1975; Shelford 1963; Küchler 1964).

MAINTENANCE REQUIREMENTS
OF CRUDE PROTEIN AND PHOSPHORUS

Neumann and Snapp (1969), Sinclair (1974d), and Peden (1972) provide useful discussions of the principal aspects of forage quality that affect nutrition in large ruminants such as cattle, the African buffalo (<u>Syncerus caffer</u>), and the American bison. These animals require a certain rate of protein assimilation in order to remain above the level of minimum maintenance (i.e., not to lose weight). This rate is dependent on the time it takes forage to pass through the rumen, which in turn is conditioned by the size of food particles (Hungate 1975:39). The mastication necessary to reduce grasses to small enough particles increases with the fiber content of the forage (Hungate 1975; Sinclair 1974d: 292). Moreover, the total crude protein content of high-fiber forages generally is low. Ruminants cannot compensate to any major extent for low protein levels in forage by increasing their total intake, because they are limited by the rate at which the food can be masticated. Moreover, at low levels of forage protein microbial action in the rumen is inhibited (Hungate 1975), further reducing the rate of protein assimilation.

Sinclair (1974d:295) indicates that, at crude protein levels below about 2.8%, forage digestibility drops to zero. Minimum maintenance levels of crude protein in cattle, African buffalo, bison, and other large ruminants generally fall between about 4% and 8%. Peden (1972:90) selects a value of 7% for cattle but notes that the minimum in bison is probably lower (i.e., bison can subsist on poorer forage than cattle without losing weight). Sinclair (1974d:295) uses a value of 5% for the African buffalo, a close relative of the bison; his figure will be used here. The requirements for pregnant and lactating females are about 10% higher (Neumann and Snapp 1969:182).

Minimum maintenance requirements of phosphorus, another critical nutrient for ruminants, have not been established for bison. Cattle require about 0.15% (Neumann and Snapp

1969:182). This figure will be used in subsequent discussion.

PRESENT-DAY FORAGE QUALITY
IN THE SOUTHERN PLAINS

The crude protein and phosphorus contents, expressed as percentage of total dry matter, of the principal grass species in the Southern Plains are summarized in·table 15. Average protein and phosphorus values, determined for each major phenological or growth stage (e.g., immature, early bloom, mature), are presented in table 16 and plotted in figure 36 against the average monthly precipitation recorded in Roswell for 1931-60 (Pieper et al. 1978; Houghton 1974). Also shown in the figure are the approximate timing in the Southern Plains of the rut (Halloran and Glass 1959:369) and the calving season (Halloran 1968; Halloran and Glass 1959), and the normal duration of lactation in bison (McHugh 1958: 34).

It is clear from figure 36 that bison in the Southern Plains would have been operating at or below maintenance levels through most of the winter and spring. Average crude protein levels drop below 5% by November and remain at submaintenance values until about May. Protein values of blue grama, the most important forage species in the region, drop to about 3.3% in the winter and early spring, only a few tenths of a percent above the theoretical zero-digestibility level (Pieper et al. 1978:22; Sinclair 1974d: 295).

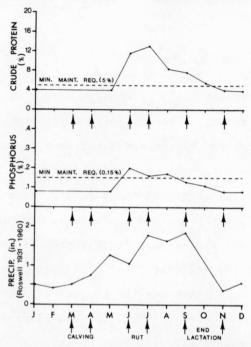

Fig. 36. Seasonal changes in average crude protein and phosphorus levels of major Southern Plains grasses in relation
to minimum maintenance requirements of bison (seasonal
fluctuations in local precipitation also shown).

Under actual grazing conditions, protein shortages may

have been more acute than these data indicate. The crude

protein values summarized in table 16 are averages for whole

plants (aerial parts). But protein is not uniformly dis-

tributed throughout the plant; instead, the highest levels

are found in the leaves (Sinclair 1974d). Thus, once new

growth ceases after the summer rains and the grasses enter

their dormant phase, the first parts to be grazed off by

Table 15

Crude protein and phosphorus levels (%) of major
Southern Plains grass species at different stages of growth

Grass[1]	Season[2]	Growth Stage	Crude Protein[3] (%)	Phosphorus[3] (%)
Bluestem, big (Andropogon gerardi)	Warm	Immature	15.6	0.21
		Early bloom	10.1	0.17
		Midbloom	--	--
		Full bloom	6.6	0.10
		Mature	4.9	--
		Overripe	2.9	0.11
		Dormant	2.4[6]	0.04[6]
Bluestem, little (Andropogon scoparius)	Warm	Immature	12.0[4]	0.13
		Early bloom	16.0	0.11
		Midbloom	7.2	--
		Full bloom	10.0[4]	0.09
		Mature	8.4[4]	0.08
		Overripe	--	0.07
		Dormant	5.0[4]	0.11[4]
Bluestem, sand (Andropogon hallii)	Warm	Immature	11.5	--
		Early bloom	--	--
		Midbloom	--	0.20
		Full bloom	5.2	0.11
		Mature	3.7	0.11
		Overripe	2.0	0.04
		Dormant	--	--
Bluestem, silver (Andropogon saccharoides)	Warm	Immature	11.8	0.22
		Early bloom	--	--
		Midbloom	6.6	0.17
		Full bloom	5.7	0.12
		Mature	3.5	0.08
		Overripe	--	--
		Dormant	2.5[10]	--
Bristle grass, plains (Setaria macrostachya)	Warm	Immature	17.5	0.23
		Early bloom	10.3	0.20
		Midbloom	--	--
		Full bloom	--	--
		Mature	--	--
		Overripe	--	--
		Dormant	--	--

Table 15 (cont.)

Grass[1]	Season[2]	Growth Stage	Crude Protein[3] (%)	Phosphorus[3] (%)
Buffalo grass (Buchloe dactyloides)	Warm	Immature	6.9[4]	0.23
		Early bloom	10.6	---
		Midbloom	---	---
		Full bloom	9.8	---
		Mature	5.9	---
		Overripe	---	0.16
		Dormant	5.7[4]	---
Burro grass (Scleropogon brevifolius)	Warm	Immature	---	---
		Early bloom	---	---
		Midbloom	---	---
		Full bloom	---	---
		Mature	---	---
		Overripe	---	---
		Dormant	---	---
Dropseed, mesa (Sporobolus flexuosus)	Warm	Immature	---	---
		Early bloom	17.5	---
		Midbloom	8.9	0.17
		Full bloom	10.8	0.15
		Mature	7.6	0.12
		Overripe	4.5	0.06
		Dormant	<4.8[9]	---
Dropseed, sand (Sporobolus cryptandrus)	Warm	Immature	11.8[5]	0.28[5]
		Early bloom	12.2	0.14
		Midbloom	---	---
		Full bloom	7.3[4]	---
		Mature	5.7	0.13
		Overripe	5.8	0.07
		Dormant	4.5[5]	0.06
Grama, black (Bouteloua eriopoda)	Warm	Immature	11.5	0.16
		Early bloom	9.3	0.16
		Midbloom	---	---
		Full bloom	---	---
		Mature	6.4	0.11
		Overripe	5.2	0.06
		Dormant	<4.9[4][9]	---

Table 15 (cont.)

Grass[1]	Season[2]	Growth Stage	Crude Protein[3] (%)	Phosphorus[3] (%)
Grama, blue (Bouteloua gracilis)	Warm	Immature	11.5	0.16
		Early bloom	11.3	---
		Midbloom	---	---
		Full bloom	---	---
		Mature	7.9	0.07
		Overripe	---	---
		Dormant	3.3[5]	0.08[5]
Grama, hairy (Bouteloua hirsuta)	Warm	Immature	---	0.14
		Early bloom	---	---
		Midbloom	---	---
		Full bloom	---	---
		Mature	5.6	0.08
		Overripe	---	---
		Dormant	---	---
Grama, side oats (Bouteloua curtipendula)	Warm	Immature	11.6	0.18
		Early bloom	---	---
		Midbloom	8.4	0.12
		Full bloom	7.1	0.10
		Mature	4.7	0.08
		Overripe	3.0	0.07
		Dormant	2.5[5]	0.07
Muhly, bush (Muhlenbergia porteri)	Warm	Immature	---	---
		Early bloom	---	---
		Midbloom	---	---
		Full bloom	---	---
		Mature	---	---
		Overripe	7.3	0.09
		Dormant	---	---
Needle and thread (Stipa comata)	Cool	Immature	12.0	0.16
		Early bloom	11.4	---
		Midbloom	8.9	---
		Full bloom	6.6[4]	0.11
		Mature	4.7[4]	0.09
		Overripe	2.9	0.07
		Dormant	4.0[4][8]	0.07

Table 15 (cont.)

Grass[1]	Season[2]	Growth Stage	Crude Protein[3] (%)	Phosphorus[3] (%)
New Mexico feather grass (Stipa neomexicana)	Cool	Immature	---	---
		Early bloom	---	---
		Midbloom	---	---
		Full bloom	---	---
		Mature	---	---
		Overripe	---	---
		Dormant	---	---
Sacaton, alkali (Sporobolus airoides)	Warm	Immature	8.8[4]	0.24[4]
		Early bloom	---	---
		Midbloom	---	---
		Full bloom	7.6[4]	0.22[4]
		Mature	6.2[4]	0.14[4]
		Overripe	---	---
		Dormant	3.5[4]	0.08[4]
Three-awn (Aristida spp.)	Warm	Immature	9.2	0.14
		Early bloom	---	---
		Midbloom	6.1	0.10
		Full bloom	4.9	0.08
		Mature	3.2	0.08
		Overripe	---	0.06
		Dormant	---	---
Tobosa (Hilaria mutica)	Warm	Immature	10.6	0.20
		Early bloom	---	---
		Midbloom	---	---
		Full bloom	6.4	0.12
		Mature	5.8	---
		Overripe	3.9	---
		Dormant	<4.0[7] [9]	0.07[7]
Vine mesquite (Panicum obtusum)	Warm	Immature	12.6	0.25
		Early bloom	13.5[5]	---
		Midbloom	---	---
		Full bloom	10.2[5]	0.19
		Mature	5.4	---
		Overripe	4.6[5]	---
		Dormant	4.1[5]	0.11[5]

Table 15 (cont.)

Grass[1]	Season[2]	Growth Stage	Crude Protein[3] (%)	Phosphorus[3] (%)
Windmill grass, hooded (Chloris cucullata)	Warm	Immature	11.9	0.23
		Early bloom	---	---
		Midbloom	---	---
		Full bloom	0.17	0.17
		Mature	---	0.16
		Overripe	---	---
		Dormant	---	---
Windmill grass, tumble (Chloris verticillata)	Warm	Immature	12.4	0.20
		Early bloom	---	---
		Midbloom	---	0.20
		Full bloom	8.6	0.16
		Mature	5.8	0.12
		Overripe	---	---
		Dormant	5.0[10]	---

[1]Species list for southeastern New Mexico (Bureau of Land Management 1979).
[2]Cool-season species undergo maximum growth in spring or early summer; warm-season grasses attain peak growth in mid- to late summer (assignments based on Bureau of Land Management 1979; Phillips Petroleum Company 1963; Colorado State University Extension Service 1981; Gay and Dwyer 1970; Redetzke and Paur 1975).
[3]Percentage of total dry matter; data obtained from National Research Council 1969, 1971, except where noted otherwise.
[4]Rodgers 1966; Cook, Child, and Larson 1977.
[5]Pieper et al. 1978.
[6]Pieper 1977.
[7]Neuenschwander, Sharrow, and Wright 1975.
[8]Wallace, Free, and Denham 1972.
[9]Paulsen and Ares 1962.
[10]Willard and Schuster 1973.

Table 16

Average crude protein and phosphorus levels (%) of major Southern Plains grass species at different stages of growth (see table 15 for original data)

Growth Stage	Crude Protein (mean %)	Phosphorus (mean %)
Immature	11.4 (16)[1]	0.20 (16)[1]
Early bloom	12.9 (10)	0.16 (5)
Midbloom	8.4 (6)	0.17 (7)
Full bloom	7.7 (14)	0.13 (11)
Mature	5.7 (17)	0.11 (15)
Overripe	4.1 (12)	0.08 (12)
Dormant	4.0 (14)	0.08 (9)

[1]Values in parentheses indicate number of cases averaged.

large ruminants are the highly nutritious tops. By late spring the remaining forage is significantly lower in crude protein content.

This recurrent seasonal protein shortage is by no means unique to the Southern Plains (for interesting general discussions concerning the nutritional inadequacy of "tropical" or "warm season" grasses such as those found in the Southern Plains, see Caswell et al. 1973 and Minson and McLeod 1970). Sinclair (1974a,b,c,d, 1975, 1977), for example, notes that the African buffalo and other ungulates in the Serengeti grasslands of Tanzania subsist on diets substantially below minimum maintenance requirements throughout most of the dry season.

Sinclair's investigations suggest that the protein available from the green component of the herbage during the dormant season may be a more critical limiting resource than total energy (Sinclair 1975). It is therefore the quality of the forage during the dry season (or winter in the case of the Southern Plains), not the total quantity or biomass of forage over the year, that plays the major role in determining the nutritional status of the animals and ultimately the size of the grazing populations. Interestingly, according to Sinclair (1975:515), consumption of large quantities of suboptimal forage reduces "the rate of utilization of the animal's fat reserves to compensate for food shortage, and hence causes a time lag in the response of the population through mortality to reduction in resources." Thus, in the Southern Plains the major period of fat depletion, particularly of the marrow bones, would be expected to occur late in the period of nutritional stress; in other words, during spring. It is also of interest that black grama, unlike most other grasses in the region, stays green during the dormant phase (Paulsen and Ares 1962). This may have been an important factor attracting bison to the more xeric parts of the Southern Plains in the winter and spring.

Phosphorus levels in Southern Plains forages also fall below minimum maintenance requirements, for even longer than crude protein (September through May). Neumann and Snapp (1969:142) note that forages from virtually all parts of New Mexico and the Texas Panhandle are deficient in phosphorus (see also Knox, Benner, and Watkins 1941). Phosphorus is critical for normal microbial activity in the rumen and for adequate milk production in lactating cows (Neumann and Snapp 1969:142-43).

The most critical period for calving females is during the last month of gestation, when most fetal growth occurs, and the first three months of lactation (Neumann and Snapp 1969:178; Robbins and Robbins 1979). Food requirements for these females may be three to four times higher than for cows without young (Sinclair 1975:503). Adequate nutrition during this four-month period is essential to producing a healthy calf and is important in determining whether the cow will conceive again in the next breeding season (Neumann and Snapp 1969:178).

It is evident from figure 36 that forages in the Southern Plains region are nutritionally inadequate for many months before the calving season. Moreover, cows that give birth early in the season may be lactating for up to three months while subsisting on a submaintenance diet. It is clear from these data that weight losses in females, including depletion of stored fat reserves, may be substantial by late spring (cf. Sinclair 1974d:302; Halloran 1968:24; see also Rice et al. 1974). Neumann and Snapp (1969:221) note that weight losses of between 150 and 200 lbs (ca. 70-90 kg) are common in calving female cattle wintered on open range. This amounts to a loss of 15% to 20% of their normal body weight. The average figure in bison would probably be somewhat smaller. These data may help account for the observation by Halloran (1968) that Southern Plains bison cows calve, on the average, only two out of every three years.

PREHISTORIC RANGE CONDITIONS
IN THE SOUTHERN PLAINS

It is clear from the discussion thus far that pregnant or lactating bison cows would have faced serious nutritional shortages each spring in the Southern Plains. Thus, discrimination against females during the butchering and processing of carcasses at Garnsey may simply reflect the expected stressful conditions of spring and may offer no insights into the general nature of environmental conditions in the region at other seasons. There are several lines of evidence, however, suggesting that poor pasture conditions in the region may have been the norm throughout the year during the mid- to late fifteenth century rather than an occasional or seasonal phenomenon in an otherwise generally favorable period.

Published Paleoclimatic
Reconstructions

Direct evidence for environmental conditions in the Southern Plains during the fifteenth century is at present very limited. Nevertheless, paleoclimatic reconstructions based on modern climate analogues by Baerreis and Bryson (1966:114), Bryson, Baerreis, and Wendland (1970:68-69; see also Duffield 1970:265) and by Sanchez and Kutzbach (1974)

do point to increasing aridity in the region after about
A.D. 1450 or 1500.

The reconstructions by Sanchez and Kutzbach (1974), if
verified by future research, are of particular interest because they indicate the potential complexity of paleoclimatic conditions in the region. First, their study suggests
that only the more southerly portions of the Southern Plains
may have become more arid; conditions may have improved in
nearby areas to the north. Second, their reconstructions
suggest that in the eastern portions of the Southern Plains
a higher proportion of the annual precipitation may have
fallen in the winter, whereas in the western portions, and
in much of the Southwest, the precipitation may have become
more concentrated in the summer months. A shift toward a
winter-dominant rainfall pattern would have exacerbated already marginal forage conditions for bison wintering in the
Southern Plains by reducing moisture during the critical
growth period of the grasses and by leaching nutrients from
the forage during the dormant phase (Paulsen and Ares 1962).

Published tree-ring sequences from archeological sites
in the Southwest at or close to the western margins of the
Plains also point to generally increased aridity in the fifteenth century. Sites providing useful sequences include
Arroyo Hondo Pueblo near Santa Fe, New Mexico (Rose, Dean,
and Robinson 1981), Tijeras Pueblo near Albuquerque, New
Mexico (Cordell 1980), and Gran Quivira Pueblo south of Al-

buquerque (Dean and Robinson 1977). The Arroyo Hondo and
Gran Quivira sequences may also point to greater variability
in precipitation, a pattern that affected more northerly
latitudes as well (see, e.g., Reher and Frison 1980:55 and
discussion in Lamb 1977:465).

Dated pollen sequences from the Southern Plains span-
ning the period of interest to us here are not yet available. Ongoing palynological investigations in the Garnsey
Wash, however, may soon provide a clearer picture of prehistoric local and regional climatic and environmental conditions (Stephen A. Hall, personal communication).

Tooth Wear

Perhaps the most direct line of evidence at present
available for suboptimal range conditions in the Southern
Plains during the fifteenth century concerns the attrition
of the lower and upper molars of the Garnsey bison. Wilson
(1980:98-99, 106) found that the average attrition rate was
significantly higher than the highest rates observed by
Reher (1974:120) and by Todd and Hofman (1978:69) at kill
sites in the Northern Plains (Casper and Horner, respectively). Reher (1974:120) felt that the high rate of attrition
in the Casper sample indicated a bison population under
stress. The even higher values at Garnsey, therefore, may
reflect a bison population subsisting on extremely marginal
forage (Wilson 1980:125).

Wilson combined the dentitions from all stratigraphic levels at Garnsey in order to work with the largest possible sample. When the wear rates are determined separately by stratigraphic unit (units A and B), a clear directional trend becomes apparent (table 17; mandibles only; unit A maxillary sample inadequate). The rate of attrition of all three lower molars increases sharply from unit B to unit A, pointing toward progressively degenerating range conditions during the mid- to late fifteenth century. Unfortunately, the sample of sexed postcranial remains from unit A is too small to determine whether there was a parallel increase in discrimination against calving females in the butchering process as pasture conditions deteriorated.

Table 17

Attrition rate of adult bison lower molars from Garnsey Site

Lower Molar	Unit B (mm/yr)	Unit A (mm/yr)
M_1	2.94 (6)[1]	4.33 (5)[1]
M_2	5.63 (6)	6.27 (5)
M_3	6.10 (6)	6.64 (6)

[1] Values in parentheses indicate sample size.

The accelerating rate of attrition of the Garnsey dentitions during the latter half of the fifteenth century probably reflects a combination of increasing forage abrasiveness and more grit in the diet. Forage abrasiveness is due largely to minute particles of plant opal or silica

(phytoliths). These particles, which are especially abundant in grasses, are harder than tooth enamel and have been
widely recognized as a major cause of tooth wear in grazers
(Baker, Jones, and Wardrop 1959; Walker, Hoeck, and Perez
1978; Covert and Kay 1981).

Numerous studies have shown that silica tends to con-
centrate in the aerial parts of grasses, especially in the
leaves and inflorescences of mature plants (Handreck and
Jones 1968:456). A wide variety of factors influence the
amount of silica deposited in the leaves, including plant
species, availability of silica in the soil, soil pH, presence and abundance of various soil nutrients, and especially
rate of transpiration (Jones and Handreck 1967; Dore 1960;
Handreck and Jones 1968; Lewin and Reimann 1969; Johnston,
Bezeau, and Smoliak 1967).

In general, grasses growing under warmer and more arid
conditions, and subject to higher rates of water loss
through transpiration, develop higher concentrations of silica than grasses growing in cooler or moister situations
(Dore 1960; Johnston, Bezeau, and Smoliak 1967). Thus a
shift toward greater aridity in the Southern Plains, which
would have favored an increase in the proportion of drought-
resistant grasses such as blue grama, would probably have
led to higher overall concentrations of silica in the forage
(Johnston, Bezeau, and Smoliak 1967; Boggino 1970). This in
turn would have accelerated tooth wear in the bison. Low

phosphorus levels, which are characteristic of soils throughout New Mexico and the Texas Panhandle, also favor elevated levels of silica in grasses, and hence higher rates of tooth wear (Jones and Handreck 1967).

Grit consumed along with the forage also plays a major, if not dominant, role in the tooth wear of ruminants such as cattle and bison. Healy (1973; also 1968, cited in Spedding 1971:112), for example, reported that cattle may ingest up to 450 kg of soil per animal per year. In addition, Johnston, Bezeau, and Smoliak (1967) observed that large quantities of silica dust adhere to grasses, particularly after the summer rainy season has ended.

Greater aridity in the Southern Plains during the fifteenth century would have favored an increase in the total amount of grit bison ingested for several reasons. First, if modern climatic analogues provide any indication of past conditions, it is very likely that drier periods were also windier (Rasmussen, Bertolin, and Almeyda 1971; Borchert 1950). Second, drought would have reduced the size of individual grass plants and decreased total basal cover. These changes would have forced bison to crop closer to the ground and would have exposed larger areas of the soil to wind erosion (Buffington and Herbel 1965; Coupland 1958; Francis and Campion 1972; Herbel, Ares, and Wright 1972; Humphrey 1958, 1962; Nelson 1934; Paulsen and Ares 1962; Sims and Singh 1978; Valentine 1970; Van Dyne 1973, 1975).

Finally, a reduction in total rainfall would have permitted more dust to adhere to the grasses (Johnston, Bezeau, and Smoliak 1967).

Thus the accelerating tooth wear observed at the Garnsey Site probably indicates reduced effective moisture in the region during the latter half of the fifteenth century, owing either to decreased total rainfall or to a shift in the seasonal timing of precipitation relative to the growth season of the principal forage species, or perhaps to increased effective temperatures (but see Sanchez and Kutzbach 1974; LaMarche 1974; and Lamb 1977 concerning projected cooler temperatures during the period of concern).

Carbon Isotope Chemistry and Bison Diet

Shifts in effective moisture directly affect the quality, quantity, and composition of the forage available to grazers such as bison and lead to changes in the composition of the animals' diets (see Van Dyne 1975 and references therein). Some of these changes are directly reflected in the carbon isotope chemistry of the bones.

The grasses in the Southern Plains fall into two broad groups on the basis of their seasonal growth patterns. The majority of species (generally 70-80% or more) are warm season types, which undergo maximum growth during the period of peak temperatures in mid- to late summer (Teeri and Stowe 1976). The minority are cool season forms, which reach maximum growth earlier in the year, normally in spring or early summer (Colorado State University Extension Service 1981; Phillips Petroleum Company 1963; Gay and Dwyer 1970; Van Dyne 1973). Most warm and cool season species have different photosynthetic pathways of carbon fixation. The former are characterized by the so-called Hatch-Slack or C_4 pathway, whereas the latter are characterized by the Calvin or C_3 pathway (Bender 1968; Doliner and Jolliffe 1979; Downton 1975; Eickmeier 1978; Hatch, Slack, and Johnson 1967; Misra and Singh 1978; Syvertsen et al. 1976; Williams and Markley 1973).

These two carbon pathways differ in the degree to which they fractionate against the heavier isotopes of carbon. Thus, collagen from bones of grazers with a diet consisting exclusively of C_4 plants has ^{13}C fractionation values of about -6.5 per mil, compared with values of about -21.5 per mil in bones of animals with a nearly pure C_3 diet (Sullivan and Krueger 1981; Land, Lundelius, and Valastro 1980; van der Merwe and Vogel 1978; van der Merwe, Roosevelt, and Vogel 1981; Tauber 1981; DeNiro and Epstein 1978a,b). The ratio of the stable carbon isotopes, ^{12}C and ^{13}C, in the bones of bison therefore offers a direct means of estimating the proportional contribution to the diet of warm and cool season grasses.

Sullivan and Krueger (1981) analyzed the $^{13}C/^{12}C$ ratio of bones from a series of modern herbivores with known diets. They found that the dietary contribution of C_4 plants could be estimated directly from the isotope ratios of both the apatite and the gelatin fractions of the bone. In their study they also analyzed bones from several prehistoric sites, including two specimens from level B3 of the Garnsey Site (GX-5685 and GX-5681; their sample numbers 17 and 18; see table 6 above).

Their analyses indicated that the diet of level B3 bison was composed of about 85% C_4 plants (Sullivan and Krueger 1981:334). This value is very close to the figure of about 80% warm season grasses noted by Peden (1972:39, 1976) in the diet of modern fistulated bison on the Pawnee Grassland experimental range in northeastern Colorado. Peden's study showed that, on lightly grazed pastures, C_3 grasses were important primarily during the spring, when they composed nearly 54% of the diet. Of particular interest to us here, however, is that Peden observed a sharp increase in the spring contribution of cool season grasses to nearly 80% of the diet under conditions of heavy grazing. He attributes this dietary shift in the spring to the fact that by the end of winter the bison have grazed off the warm season species.

These interesting observations lead to the expectation that the deteriorating range conditions, evidenced in level A2 at Garnsey by accelerating wear of the bison dentitions, may also have been accompanied by a shift, particularly during the spring, toward more cool season (C_3) plants in the diet of the level A2 bison. Such a dietary change would in turn be accompanied by a shift toward more negative average ^{13}C fractionation values in the bones of the animals.

To test this expectation, bone samples from level A2 were submitted to Geochron Laboratories for isotope analysis (I am extremely grateful to Harold W. Krueger and Charles H. Sullivan for undertaking these additional analyses and for permitting me to use these data as soon as they became available). The results are summarized in table 18. The trend in the isotope values is relatively clear-cut. Combining all unit B values in tables 6 and 18, the average ^{13}C fractionation values for apatite and gelatin are -0.2 and -9.0 per mil, respectively. As predicted, both unit A figures are more negative than their unit B counterparts (average apatite, -1.0 per mil; average gelatin, -12.5 per mil).

These results point toward an increase in the contribution of cool season species to the diet of the bison in the younger levels at Garnsey. It is interesting in this regard that the grass culms from unit A submitted to Beta Analytic for radiocarbon dating (Beta-1925; see table 6 above), while resembling sacaton (Sporobolus airoides; a warm season grass; Richard I. Ford, personal communication), must be a cool season form on the basis of the ^{13}C fractionation value

(-24.2 per mil). This grass may have been a major factor attracting bison to the Garnsey Wash during the spring.

Table 18

^{13}C fractionation analyses (apatite and gelatin) of bison bones from level A2 and level B4 of the Garnsey Site (see table 6 above for additional fractionation values)

Provenience	Laboratory Number[1]	Element	Level	^{13}C Fractionation[3]	
				Apatite	Gelatin
504S507W/15	CR-17263	Mc	B4	-1.4(2)	-11.3(3)
517S493W/2	CR-17264	Mt	(A2)[2]	-2.6(2)	-15.4(3)
511S476W/14	CR-17265	R	A2	-0.4(6)	-12.6(4)
518S485W/1	CR-17266	Mc	A2	-0.8(2)	-12.0(3)
518S485W/3	CR-17267	R	A2	-0.7(2)	-11.8(2)
518S485W/4	CR-17268	U	A2	-0.3·	-10.5

[1]Carbon isotope analyses by Harold W. Krueger and Charles H. Sullivan, Geochron Laboratories, Cambridge, Mass.
[2]Specimen recovered from interface of levels A2 and B3, making precise stratigraphic placement difficult; condition of bone more closely resembles that of other level A2 materials.
[3]Figures in parentheses indicate number of replicate determinations averaged by lab; fractionation values (per mil) measured relative to PDB standard.

Life Expectancy
of the Garnsey Bison

Finally, there is one additional line of evidence that points toward widespread suboptimal pasture conditions in southeastern New Mexico. This is the life expectancy of the Garnsey bison, a parameter that obviously is closely linked to dental attrition (see, e.g., Sinclair 1974d:302). Predictably, the life expectancy of the Garnsey bison was short (about twelve years). This figure is similar to the values

observed at the Casper Site in Wyoming (Reher 1974), which
also displayed extremely high rates of dental attrition.
The life expectancy of bison from kills displaying lower
rates of tooth wear (Wardell and Glenrock) was considerably
greater (ca. 16.4 and 14.5 years, respectively; Wilson
1980:100).

7. Fats and Human Hunting Strategies

INTRODUCTION

The importance to hunters of bulk differences between different prey species, and between different anatomical parts in carcasses of the same species, has been recognized for many years by archeologists, though it is only recently that attempts have been made to develop objective criteria for modeling bulk-related decision processes (Binford 1978).

The importance to hunters of an animal's physiological state or condition, on the other hand, has been considered almost solely in terms of which animals were likely to be hunted; its influence on subsequent processing has been all but ignored. Thus most prehistorians concur that seasonal differences in the condition of males and females favored preferential procurement of bulls in the spring and cows in the fall and winter. However, most prehistorians seem also to assume that all animals, once killed, would have been handled in more or less the same manner--that anatomical parts of both sexes would have been removed from the kill in proportions similar to their representation in the original kill population. This assumption is implicit, for example, in the fact that estimates of the sex ratio of bison kill populations commonly are based on the ratio obtained from one skeletal element, or at most two or three (e.g., mandible, metacarpal, or humerus).

The Garnsey case clearly demonstrates the inappropriateness of this assumption. At Garnsey the overall utility of an anatomical part appears to have been evaluated simultaneously on the basis of its bulk and its physiological condition, compared with the bulk and condition of other parts in carcasses of the same sex and with the bulk and condition of the same part in carcasses of the opposite sex.

In previous discussion of the physiological condition of the Garnsey bison and its bearing on prehistoric procurement and processing decisions at the site, I have made two basic assumptions that require more explicit consideration. First, I have assumed that changes in forage quality were reflected in broadly parallel changes in overall animal condition, especially in levels of stored body fat. Second, and of most immediate concern to us here, I have assumed that the fat levels in an animal were of more than incidental importance to the hunters. These two assumptions will now be examined more closely.

ENVIRONMENTAL CONDITIONS,
ANIMAL PHYSIOLOGY, AND FAT LEVELS

There is a voluminous literature documenting the close
relationship between environmental quality and the nutritional status or well-being of ruminants and other animals
(see references in Pond 1978; see also Riney 1955; Harris
1945; Anderson, Medin, and Bowden 1972; Ransom 1965). These
studies also show that the quantity and distribution of fat
deposits in an animal provide an extremely sensitive index
of its condition and nutritional state. Riney (1955:431),
for example, observes that "fat can be taken as a direct
measure of the condition, reflecting the metabolic level or
goodness of physiological adjustment of an animal with its
environment." Keys and Brozek (cited in Pond 1978:519) also
emphasize the importance of body fat as a measure of condition. "Interstitial and depot fat is by far the most variable constituent of the body and the one most clearly related to nutrition." Finally, Martin (1977:631) points out
the close correspondence between fat reserves and an
animal's general level of well-being. "It is a well established procedure to use the size of fat reserves as an
indicator of an animal's well-being, its general plane of
nutrition, and the favourability of its environment." These
and numerous other studies clearly testify to the close general relationship between environmental factors, such as
forage quality, and the quantity of body fat deposited in an

animal's carcass. It remains to be demonstrated, however, that the level of body fat in a carcass was of sufficient concern to prehistoric hunters to have entered significantly into their procurement and processing decisions.

ANIMAL FAT AND HUMAN DIET

Cultural Preferences for Fat-Rich Foods

Jochim (1976, 1981), on the basis of an extensive survey of the ethnographic literature, has argued that the fat content of foods is an important factor underlying the food preferences of hunters and gatherers throughout the world. He states that "high-quality protein is a requirement for survival, but since many peoples can give reasons for their food preferences, and since these reasons do not include the protein content of different foods, it becomes necessary to explore the link between stated preferences and nutritional consequences that include protein intake. One possible link between the two is the suggestion that, for lack of a better name, may be called the 'Fat Hypothesis'" (Jochim 1981:81). A few ethnographic examples will serve to illustrate the widespread preference for fatty meat among hunting populations (see Jochim 1981:81 for additional references).

> The underground species are highly desired
> because they are very fat, and animal fat
> is one of the elements most scarce in the
> Bushman diet. (Kalahari San, Botswana; Lee
> 1972:344)

All meat is classified in terms of whether
it is fat (preferred) or lean, sweet or
rank, clean or dirty, and whether it is
from a "she" or "he" animal. (Miskito,
Nicaragua; Nietschmann 1973:105)

Hunters generally seek large, fat animals
(kayambo), although in times of hunger
probably any susceptible one may be killed.
(Bisa, Zambia; Marks 1976:105)

The Pitjandjara consider the best meats to
be Kangaroo and Euro, making little distinction between them. They are selective
with such animals. When killed they immediately feel the body for evidence of the
presence of caul fat. If the animal is
njuka, or fatless, it is usually left,
unless they are themselves starving.
(Pitjandjara, Australia; Tindale 1972:248)

Bear are greatly desired as food, especially because of the large amount of fat
they yield. They are taken at every oppor-
tunity, but are eagerly sought during the
fall when they are fattest. . . . Beaver
rival bear as an item relished in the diet
because of their high fat content. (Mistassini Cree, Canada; Rogers 1972:111-12)

Fat porcupine, bear and beaver meat are
esteemed, particularly the tail of the
latter. They are fond of marrow and fat.
(Hidatsa, North American Plains; Matthews
1877:24)

A final example, also from the Hidatsa, is of special

relevance to the discussion of the Garnsey Site, because it

describes one means by which hunters could identify a bull

bison with adequate fat reserves in the spring.

"Now let me tell you again how to judge
if one of the bulls is fat. As you come
close, observe if the hair along the spine
and just back of the eyes is black. Those
so marked are the fat ones." The reason
of this is that the black hair marked where

the buffalo had begun to shed his hair.
Under the black spots were layers of fat
that in those places made the buffalo shed
a little earlier than his leaner fellows.
But such a sign was of value only in the
spring and was found only on bulls, not on
cows. (Wilson 1924:306)

Biological Importance of Fat

Jochim (1981:81) points out that "emic" preferences among hunter-gatherers for fatty foods often represent biologically sound food choices. Thus, fat provides the most concentrated source of energy in the diet, supplying over twice as many calories per gram as either protein or carbohydrate (Guthrie 1975:45). In addition, dietary fat is a source of essential fatty acids; it provides a vehicle for the fat-soluble vitamins A, D, E, and K; and it is important in the development and functioning of cell membranes (Food and Agriculture Organization 1977; Pond 1978; Allen et al. 1976; Guthrie 1975:45-47). Finally, fat makes foods taste better (Food and Agriculture Organization 1977), and many fat-rich foods also tend to be high in protein (Jochim 1981:82-83).

Although dietary fat serves a variety of functions, it is most important as a concentrated energy source. Obviously, however, energy is not derived only from fat; carbohydrate is another extremely important source, and protein beyond basic requirements also provides calories (Guthrie 1975; Taylor and Pye 1966; Chaney and Ross 1971). Given our

focus on the food-getting strategies of hunters, it is the
energy derived from hunted foods that most concerns us here.
Put another way, we are interested in the viability of a
diet composed largely, if not entirely, of meat; and, more
specifically, we are concerned with the importance in an
all-meat diet of fatty versus lean meat.

Eskimos are well known for diets that, aboriginally,
were largely, if not totally, devoid of plant foods (cf.
Draper 1977). In a widely publicized experiment, two arctic
explorers, Stefansson and Andersen, sustained themselves
under close medical supervision for an entire year on an
all-meat (including fat) diet (Stefansson 1935a,b, 1936; see
also Draper 1977; Chaney and Ross 1971:114-15). This experiment demonstrated the viability in a nonarctic context
of a diet without plant foods. Neither of the participants
developed scurvy, despite the exceedingly low levels of
vitamin C in meat (Watt and Merrill 1963; Draper 1977:310).
Stefansson (1936:183) maintained that "if you have some
fresh meat in your diet every day, and don't overcook it,
there will be enough C from that source alone to prevent
scurvy." Recent studies suggest that bone marrow rather
than meat may have been their major source of vitamin C;
marrow appears to be a repository for significant quantities
of this important nutrient (Field 1976:68).

While an all-meat diet may be viable, Stefansson has
stressed that a diet based entirely on lean meat would

quickly lead to nutritional disorders, and eventually to death. He states emphatically that "fat is the most important ingredient of an Arctic ration" (Stefansson 1944:221). He goes on to describe in detail the deleterious effects of subsisting on the lean meat of rabbits in northern latitudes. These animals are notoriously low in body fat, and relying on them for food leads to a condition known as "rabbit starvation."

> If you are transferred suddenly from a diet normal in fat to one consisting wholly of rabbit you eat bigger and bigger meals for the first few days until at the end of about a week you are eating in pounds three or four times as much as you were at the beginning of the week. By that time you are showing both signs of starvation and of protein poisoning. You eat numerous meals; you feel hungry at the end of each; you are in discomfort through distention of the stomach with much food and you begin to feel a vague restlessness. Diarrhoea will start in from a week to 10 days and will not be relieved unless you secure fat. Death will result after several weeks. (Stefansson 1944:234)

Observations similar to Stefansson's are common in the ethnographic literature. The following examples further illustrate the inadequacy, in northern latitudes, of a diet based on lean meat.

> The quantity of food consumed at one meal naturally varies according to the amount of fat it contains. Men and dogs will half-starve on a diet of lean caribou-meat, however plentiful, whereas half the quantity of blubbery seal-meat will satisfy their desires and keep them

well nourished. (Copper Eskimo, Canada;
Jenness 1923:100)

If people had only rabbits at such times
they would probably starve to death, because these animals are too lean. The same
might be true if they could get only thin
moose. People cannot live on lean meat
alone, but if they have enough fat they
can survive indefinitely. (Kutchin, Alaska;
Nelson 1973:142)

There are many similar observations in the ethnohistoric literature, and these are by no means confined to the
arctic. For example, in June 1830 Warren A. Ferris, during
his expedition to the Rocky Mountains, made the following
observations concerning the poor sustenance provided by lean
buffalo.

We killed here a great many buffalo, which
were all in good condition, and feasted,
as may be supposed, luxuriously upon the
delicate tongues, rich humps, fat roasts,
and savoury steaks of this noble and
excellent species of game. Heretofore
we had found the meat of the poor buffalo
the worst diet imaginable, and in fact
grew meagre and gaunt in the midst of
plenty and profusion. But in proportion
as they became fat, we grew strong and
hearty. (Phillips 1940:42)

Jedediah Smith led a group of trappers along the Klamath
River in California in May 1828. According to Dale Morgan
(cited in Allen 1979:457),

the party made only 3 miles on the
18th, which taxed their strength to
the utmost. "The men were almost as
weak as the horses, for the poor
[spring] venison of this country
contained little nourishment."

Finally, Randolph B. Marcy in the winter of 1857-58 provided an extremely illuminating comment regarding the inadequacy of a diet based on fat-depleted meat.

> We tried the meat of horse, colt, and
> mules, all of which were in a starved
> condition, and of course not very tender,
> juicy, or nutritious. We consumed the
> enormous amount of from five to six
> pounds of this meat per man daily, but
> continued to grow weak and thin, until,
> at the expiration of twelve days, we
> were able to perform but little labor,
> and were continually craving for fat
> meat. (Marcy 1863:16)

Of particular interest are frequent references to situations in which hunters avoided or abandoned animals they considered too lean in favor of fatter animals, even when the hunters themselves were short of food. A typical example was recorded by Lewis and Clark in December 1804.

> Captain Lewis went down with a party
> to hunt. They proceeded about 18 miles;
> but the buffalo having left the banks
> of the river they saw only two, which
> were so poor as not to be worth killing.
> (Coues 1893:211)

In February 1805 Lewis and Clark made a similar entry in their diary.

> Captain Clark returned last evening with
> all his hunting party. During their ex-
> cursion they had killed 40 deer, 3 buffalo,
> and 16 elk; but most of the game was too
> lean for use. (Coues 1893:233)

As a final example, Jacob Fowler, during his expedition to the Rocky Mountains, made the following comment in his journal in February 1822.

> Hunters out Early--Killed one Cow Buffe-
> low With In four Hunderd yards of Camp--
> but so Poor the meat Was not Worth Save-
> ing. (Coues 1898:97; spelling as in
> original)

The nutritional literature offers several interesting
and important insights into reasons why hunters might have
abandoned lean or fat-depleted carcasses, even when they
themselves were short of food. Two aspects of protein metabolism appear to be of particular relevance to this issue:
(1) the high "specific dynamic action" of protein ingestion;
and (2) the relatively low "protein-sparing" or "nitrogen-
sparing" effect of fat, compared with carbohydrate, in low-
calorie diets.

The specific dynamic action (SDA) of food refers to the
rise in metabolism or heat production that results from in-
gesting food (Chaney and Ross 1971:45; Barnes 1976:12-13;
Taylor and Pye 1966:39-40; Guthrie 1975:94-95; Bigwood
1972:xxii; Wing and Brown 1979:22). The SDA of a diet consisting largely of fat is about 4%, while that of a diet
high in carbohydrate is about 6%. In striking contrast, the
SDA of a diet consisting almost entirely of protein is close
to 30%; in other words, for every hundred calories of protein ingested, an extra thirty are needed to compensate for
the increase in metabolism. Thus, heavy or complete reliance on fat-depleted meat would have elevated the total energy needed to support a group of hunters. The effect of
increased energy demands owing to protein ingestion obvious-

ly would be greatest during periods when overall calorie intake was restricted.

The other aspect of protein metabolism of interest to us here is the protein-sparing effect of nonprotein dietary energy. Proteins are complex substances built up of nitrogen-containing amino acids. The amount of nitrogen present in protein is relatively constant and uniform, averaging about 16% (Guthrie 1975:53-54). Thus protein metabolism is often studied indirectly by monitoring the amounts of nitrogen ingested and excreted (Guthrie 1975: 64-66). If the amount of nitrogen lost is lower than the amount taken in, the individual is in positive nitrogen balance; that is, protein is being gained. Similarly, if the amount excreted exceeds the amount ingested, the individual is in negative nitrogen balance; protein is being lost.

Both fat and carbohydrate exert a protein- or nitrogen-sparing action on protein metabolism (Munro 1964; Richardson et al. 1979; Gelfand, Hendler, and Sherwin 1979; Peret and Jacquot 1972; Goodhart and Shils 1980:820, 1128; Sim et al. 1979). This means that when either source of energy is increased the amount of nitrogen excreted in the urine declines. Numerous studies have demonstrated, however, that the protein-sparing effect of carbohydrate is much greater than that of fat (Richardson et al. 1979; Gelfand, Hendler, and Sherwin 1979). "It has long been known that if fat and

carbohydrate are given in isocaloric amounts they do not exert an equal protein-sparing effect, and that carbohydrate
is more effective than fat in promoting the utilization of
dietary protein" (Richardson et al. 1979:2224).

One indication of the greater protein-sparing effect of
carbohydrate is that when it is administered to fasting sub-
jects there is a noticeable decline in the amount of nitrogen excreted. No improvement in nitrogen balance is obtained when fat is given to fasting individuals. Another
indication is that when fat is substituted for carbohydrate,
while holding total dietary protein and energy intake constant and at normal levels, nitrogen losses increase sharply. Munro (1964) suggested that increased nitrogen output
caused by substituting fat for carbohydrate was a transient
phenomenon lasting at most about fifteen days. More recent
studies by Richardson et al. (1979), however, indicate that
the negative nitrogen balance produced by fat substitution
may persist for much longer periods and may have a much more
detrimental effect. They observed that "on average, the net
protein utilization of . . . protein could be raised by
about 13% when the diet provided the majority of the energy
from carbohydrate" (Richardson et al. 1979:2223). Of particular interest, Richardson et al. (1979:2223) noted a considerably greater protein-sparing effect (values approaching
50%) in individuals whose caloric intake was marginal.

If these observations on the specific dynamic action of protein ingestion and the protein-sparing effect of dietary carbohydrate stand up to further scrutiny, they have important implications for understanding certain critical aspects of hunter-gatherer subsistence behavior. Let us assume in the following discussion that we are dealing with hunting populations during the late winter and spring. This is the most stressful period of the year, often characterized by diminishing supplies of stored food and the absence of collectible plant foods. To compensate for declining reserves, hunter-gatherers may be forced to increase their hunting. This in turn may increase the proportion of meat in the diet. However, late winter and spring are not stressful only to humans; ungulates such as deer and bison also may be subsisting on diets of declining nutritional worth, and their reserves of body fat may become depleted. As a consequence, hunters may be eating greater and greater quantities of lean meat. It should be borne in mind throughout this discussion that the meat of wild ungulates generally contains considerably less intramuscular fat than the meat of domestic livestock, which have been bred specifically for high degrees of marbling (cf. Draper 1977; Speth and Spielmann, n.d.).

Because of the high specific dynamic action of protein, eating more lean meat during the late winter and spring may substantially increase the energy hunter-gatherers need to maintain themselves. Moreover, with declining carbohydrate reserves and progressively lower total calorie intake, there may be a drop in nitrogen-sparing action during protein metabolism, leading to less and less effective utilization of dietary protein.

It is clear from this discussion that diets high in lean meat are a relatively inefficient source of sustenance for hunters and gatherers at times when their intake of carbohydrate and total energy is restricted. There are several responses they may make to cope with such caloric shortfall (see Speth and Spielmann, n.d.).

First, when they make a successful kill, hunters may have to eat unusually large quantities of meat, if it is lean, to meet caloric needs and maintain positive nitrogen balance. This may help account for the seemingly anomalous gorging on meat at bison kills frequently noted by early North American travelers and explorers (see discussion in Wheat 1972:108-9).

Second, hunter-gatherers may become more selective in killing fat animals and in processing fat-rich body parts. It is this behavior that we are witnessing at the spring-season Garnsey Site. Archeologists must exercise great care in distinguishing such selectivity from "waste" when food is abundant.

Third, hunters and gatherers may shift their hunting emphasis to animal species that normally maintain relatively

high levels of body fat in winter and spring. These may include, for example, beaver, various migratory waterfowl, and
others, depending of course on location (cf. Rogers 1972).

Fourth, hunters and gatherers may augment their supplies of storable fats through labor-intensive activities
such as rendering bone grease.

Finally, in light of the substantially greater protein-
sparing capacity of carbohydrate compared with fat, hunters
and gatherers may emphasize building up storable carbohydrate reserves during the fall rather than hunting, particularly where adequate supplies of fat cannot be reliably procured. Moreover, where carbohydrate-rich wild plant foods
are also not available in quantities adequate to store for
winter use, they may supplement their carbohydrate reserves
through limited cultivation or by trading with horticultural
populations (Speth and Spielmann, n.d.; conditions favoring
hunter-gatherer/horticulturalist food exchange have recently
been considered in detail by Spielmann 1982).

In the discussion above, I have focused on alternative
strategies open to hunters and gatherers to cope with occasional short-term or seasonal restrictions in their overall calorie intake. These same arguments may be extended to
situations in which climatic, environmental, demographic, or
other changes lead to long-term reductions in available energy. Under such conditions, selection may favor a permanent shift in subsistence strategies toward carbohydrate

resources. The apparent increase in reliance on plant foods in many parts of the world after the end of the Pleistocene might profitably be explored from this perspective. Similarly, a long-term increase in the availability of carbohydrate, owing, for example, to the introduction of a cultivated plant species, may alter the importance of animal fat and may lead to permanent changes in the animal species hunter-gatherers kill and the body parts they select during butchering and processing. The caloric inadequacy of a lean-meat diet and the noninterchangeability of fat and carbohydrate clearly open a number of interesting avenues of research that remain to be explored in detail.

8. Discussion and Conclusions

During the mid- to late fifteenth century, the Garnsey Site was the locus of recurrent spring-season procurement activities in which small groups of bison were surrounded or ambushed in the wash, butchered, and processed for transport to settlements that were probably to the west of the Pecos Valley. Procurement strategies were highly selective. Although cows and calves were taken, the unusually high proportion of males (60%) points to the preferential killing of bulls.

Sexing of most postcranial elements, made possible by the high degree of sexual dimorphism in bison, not only revealed great variability from element to element in the sex ratio but also showed that in many cases the original sex ratio had been reversed in favor of females. The Garnsey material therefore underscores observations in the historic and ethnographic literature that selectivity by hunters extended not only to the sex of the animals taken, but also to the manner and intensity with which the animals,

once killed, were processed. At Garnsey, what was killed
and what was subsequently left behind are diametrically opposite. Thus, as a cautionary note to archeologists, let me
emphasize that the sex ratio of most anatomical parts abandoned at a kill may bear little or no direct relation to the
sex ratio of the original kill population (see also Binford
1978:481). What remains at the site will be sharply biased
toward whichever sex was least desired at the time of the
kill. The Garnsey data also demonstrate a similar bias
against younger animals (though not necessarily against the
youngest, which are clearly underrepresented at the site).

In general, at Garnsey there was a better fit between
the sex ratio of an element and the sex ratio of the kill
population the lower the utility value of the element (i.e.,
skulls and axes provided better estimates than femurs or
humeri). But this should not be applied across the board as
a hard-and-fast rule. The atlas, for example, though very
low in general utility, nonetheless deviated sharply from
the original sex ratio, because the male element was systematically destroyed during intensive butchering of the
preferred sex. The metapodials are yet another low-utility
part whose sex ratio deviated significantly from that of the
original kill population. In this case female metapodials
were systematically rejected because their spring-season
marrow-fat reserves were depleted, whereas their male coun-

terparts at the same time of year were in better nutritional
condition.

The clear and consistent patterning of bone remains at
Garnsey in relation to the modified general utility index
indicates that processing decisions, for the most part, were
strongly conditioned by utility. The Garnsey data therefore
provide a record of decisions by prehistoric hunters concerning the selection or rejection of anatomical parts,
weighted simultaneously in terms of their utility and in
terms of specific logistic constraints and maintenance needs
of the human population at the time of the kill.

Our understanding of the principal factors entering
into the decision process needs further development and
refinement so we can make more productive use of kill-site
data to ascertain the determinants underlying prehistoric
procurement strategies. Many extremely useful comparative
insights can be gained from reanalysis of the masses of
faunal data already stockpiled in museums, combined with detailed modeling of historically and ethnographically documented Plains procurement systems.

The utility indexes developed for caribou must be extended to incorporate the tremendous degree of sexual dimorphism found in adult bison, as well as the within- and
between-sex differences in utility at different ages. The
indexes must also be expanded to deal with within- and
between-sex seasonal variations in utility. It is clear

from the literature on other ungulates that seasonal changes in condition are patterned and are therefore amenable to modeling. Although changes in the condition of males and females during the course of normal or abnormal years are broadly comparable, the changes in the two sexes differ to some extent in degree, and they are also slightly out of phase. Both sexes are in poorest shape in the spring, but males reach their lowest point earlier than females and are improving at the time when females reach their lowest point. Similarly, although both sexes improve during early summer, males improve more rapidly. In mid- to late summer, however, males decline sharply while females continue to improve gradually throughout the summer, fall, and even early winter. Both sexes decline in late winter, but males often enter this period with less fat reserve and are therefore more vulnerable to undernutrition if conditions are severe. Males, however, rebound much faster and sooner in the spring than pregnant or lactating females. Finally, females, on the average, may not calve every year (McHugh 1958; Fuller 1961; Halloran 1968; Meagher 1973). Thus their annual nutritional cycle when they are carrying fetuses will differ from that in years when they are not.

If processing decisions of hunters were closely keyed to the anatomical and physiological "facts" of bison, it is quite likely that kills that took place during seasons other than late spring will also display complex patterns of selectivity in processing. Thus, bias in the bone frequencies of kills should probably be taken as a "given," but the degree and direction of bias can be expected to vary widely depending on the age and sex structure of the original kill population, the nutritional condition of the animals at the time of the kill, and so forth. Despite the obvious complexity of the problem, a better understanding of annual patterns of variability in the utility of various anatomical parts of bison will provide more explicit sets of expectations against which to compare the archeological data.

Significant patterned deviations of the bone frequencies observed archeologically from those expected solely on the basis of utility will then provide useful insights into the nature of logistic constraints and maintenance needs of the hunters themselves. Some of the most critical factors entering into processing decisions that would become amenable to investigation include the nutritional state of the human population, time and transport constraints, and the extent to which meat was procured for storage or for immediate consumption (for a discussion of the wide range of potential factors entering into processing decisions, see Binford 1978). Given the growing interest among archeologists in the role of nutritional stress in prehistoric adaptations and culture change (cf. Minnis 1981 and references therein), careful intersite (or interlevel) comparisons of the nature and degree of selective handling, particularly of low-utility parts, may provide sensitive indicators of changing patterns of food security.

At present, given the absence of decision models developed explicitly for Plains procurement systems, the only framework available within which to evaluate the Garnsey data is the one provided by Binford (1978) for the Nunamiut Eskimo. Clearly, such long-distance comparisons are tenuous and must be made with the utmost caution. The principal intent of such comparisons at this juncture is to illustrate the as yet untapped potential of Plains kill-site data.

Comparisons with the Nunamiut case are not totally without justification. Both systems focused heavily on a single migratory species. Both relied on logistically oriented procurement strategies in which transport constraints and storage were critical. And both systems employed broadly comparable processing strategies, based heavily on utility.

Perhaps the most obvious aspect of procurement strategies at Garnsey is the apparent deliberate pursuit of males. Males provided a much greater return per animal killed, and in late spring they were probably nutritionally in better condition than females. A second important aspect of procurement strategies in evidence at Garnsey is the removal, regardless of sex, of body parts of both high and moderate utility and the discarding primarily of parts of lowest overall utility. The resultant "bulk" curves imply

that hunters were taking many parts at rates greater than
would be expected on the basis of utility alone. These curves therefore may reflect a human population subject to some
degree of food insecurity at the time of the kill events
(Binford 1978:45). The intensity of butchering at the site,
indicated by the extent of disarticulation of anatomical
units, may also point to food insecurity. On the other
hand, one could argue that the lack of selectivity toward
less-than-optimal parts reflected minimal transport constraints. This seems unlikely for two reasons. First, the
total bulk that could be removed from the site was limited
by what the hunters could transport with dogs or on their
own backs (Spielmann 1982). Second, that butchering and
processing were intensive, with few articulated units left
behind, and that bulky items such as skulls and pelvises
were systematically stripped and dumped on the site suggest
that reducing bulk was indeed an important consideration
(Binford 1978:50, 130; see also Wilson 1924:226-28, 235-36,
251-52; Wissler 1910:42).

The nature of selectivity directed specifically against
females is informative. When articulated units were recovered at Garnsey, they were for the most part from females.
They included one complete front end, several lower-limb
units, an occasional series of cervicals or thoracics, and a
few other miscellaneous sets of articulated elements. But
the overall impression is that females, once killed, under-

went relatively intensive butchering that left few articulated remains. The most striking difference in the treatment of the two sexes is the high rate of rejection of the nutritionally poorest parts of females, including necks, upper forelimbs, and especially bulky marrow bones. Again, this suggests selectivity to reduce bulk. The entire pattern at Garnsey therefore points to a procurement strategy, operating within the context of at least some food insecurity, that was directed toward optimizing the yield per animal killed while at the same time minimizing bulk.

Venturing a little further into the realm of speculation, the frequency at kill sites of phalanges and mandibles may be a useful indicator of the presence and degree of food insecurity (see Binford 1978:32, 36-38, 148-50; Dorsey 1884:293; Stefansson 1925:233). Both elements produce edible marrow and hence are potential food sources, but phalanges require considerable labor to extract useful quantities of marrow and mandibles are bulky in relation to their yield. Phalanges are problematic, however, because they often are found at processing sites and campsites with no evidence of having been broken open for marrow. They probably were commonly removed from kill sites as riders attached to more desirable limb units. Their relative scarcity at Garnsey therefore does not necessarily reflect food insecurity. Obviously, in some sites the absence of phalanges may be due entirely to taphonomic rather than cultural processes. As discussed earlier, however, this does not appear to be true for the Garnsey Site.

The mandibles, in contrast to the phalanges, are unlikely to have been removed from Garnsey as riders, since there are many more skulls and necks than mandibles left on the site. The mandible has two features that lower its desirability as a marrow bone: (1) it is bulky in relation to the amount of marrow it produces; and (2) the oleic acid content of the marrow fat is low (ca. 25%) compared with much higher values in the limbs (e.g., distal femur, ca. 51%; distal metatarsal, ca. 73%; values based on caribou; Binford 1978:24). Thus the mandibular marrow is probably less palatable than the marrow in the limbs, assuming that palatability is directly related to oleic acid content. But several characteristics make the mandible useful as an emergency food: (1) the overall proportional fat content of mandibular marrow is higher than that of limb marrow (data based on deer; Baker and Lueth 1967); and (2) although the mandibular marrow fat is mobilized sooner in undernourished animals than the marrow fat of the limbs, the percentage of fat remaining in the mandibular marrow of animals in poorest overall condition, regardless of sex, exceeds that in the limbs (Baker and Lueth 1967; Nichols and Pelton 1972, 1974; Warren 1979).

At Garnsey, nearly 70% of the expected number of mandibles had been removed from the site for processing else-

where, presumably with tongues still attached. Of the comparatively small number of mandibles left behind, approximately 25% were missing their ventral borders, pointing
toward on-site processing of marrow (there is no evidence at
Garnsey that the ventral borders had been broken off by
predator-scavengers). In sum, the comparative scarcity of
mandibles at Garnsey may point to processing within the context of relatively intense food insecurity.

The Garnsey case by itself is insufficient to discriminate clearly between occasional or seasonal food shortages
affecting only local populations and more continuous, en-
demic food insecurity in the mid- to late fifteenth century
affecting populations over a much wider area of the Southern
Plains. To the extent that by A.D. 1450 bison were a major
component of regional subsistence systems (Jelinek 1967;
Collins 1966, 1971; Dillehay 1974; Gunnerson 1972; Lynott
1980), we can speculate that food insecurity may have been
the norm rather than an infrequent or seasonal problem.
This is suggested by the relatively poor condition of the
Garnsey bison themselves, particularly as reflected by their
high rate of dental attrition and short life expectancy.

These factors, in conjunction with the carbon isotope
data, indicate that the Garnsey animals were subsisting on
extremely marginal and progressively deteriorating forage,
which in turn may have had important demographic consequences for local herds by lowering calf crops and increasing

calf and adult mortality (see discussion in Reher and Frison 1980 and in Sinclair 1974a,b,c,d, 1975). In all likelihood, bison populations in the region were extremely vulnerable to fluctuations in precipitation that altered local pasture conditions; as a consequence, their numbers, their movements, and their nutritional condition may have been highly variable and unpredictable from year to year. It seems likely that exploitation of bison in the area would of necessity have been opportunistic, with frequent recourse to other hunted, gathered, or perhaps cultivated or traded resources. Ultimately, careful comparisons among a broad range of late prehistoric settlement types in the region are needed, in conjunction with far more detailed paleoenvironmental reconstructions, in order to delineate the nature, degree, and spatiotemporal extent of food stress affecting human adaptations during this critical and fascinating period of Southern Plains prehistory.

Appendix

INTRODUCTION

To aid in sexing damaged or fragmentary limb elements,
a series of criteria have been developed using skeletal material from modern bison (<u>Bison</u> <u>bison</u>) of known sex. Comparative collections housed in the following institutions
were examined: (1) Division of Mammals, National Museum of
Natural History; (2) Department of Mammals, Museum of Comparative Zoology, Harvard University; and (3) Museum of
Zoology, University of Michigan.

MEASUREMENTS

Measurements were selected expressly to aid in sexing
mature limb elements that have suffered postdepositional
damage or that were broken prehistorically in butchering and
processing. The criteria developed here are designed to
augment existing procedures (see, e.g., Bedord 1974, 1978;
Duffield 1970, 1973; Lorrain 1968; Smiley 1979). The elements examined were the humerus, radius, metacarpal, femur,

tibia, and metatarsal. Procedures similar to those described by von den Driesch (1976) were followed in collect-
ing the data. Measurements were made with vernier calipers
and a measuring box. Values that could be determined with
relative ease and a high degree of replicability are denoted
by (+); those that were difficult to measure consistently
are denoted by (-). The terminology employed here is
derived from von den Driesch (1976), Sisson and Grossman
(see Getty 1975), Olsen (1960), and Brown and Gustafson
(1979). The procedures for obtaining each measurement are
outlined below and illustrated in figures 37-60. The raw
data are presented in tables 19-30.

Proximal humerus (figs. 37-38):

A - Depth of greater (lateral) tuberosity (+).

B - Depth of caudal (posterior) eminence of
    greater (lateral) tuberosity (+).

C - Greatest depth of proximal end. Determined
    with measuring box (+).

D - Greatest breadth of proximal end.
    Determined with measuring box (+).

E - Depth of head from cranial (anterior) edge
    of articular surface (+).

F - Depth of head from cranial (anterior) edge
    of lesser (medial) tuberosity (+).

G - Length of cranial (anterior) eminence of
    greater (lateral) tuberosity from distal

edge of rough prominence for attachment
of infraspinatus tendon (+).

H - Diagonal breadth of proximal end from
cranial (anterior) edge of lesser (medial)
tuberosity to prominent notch between head
and caudal (posterior) eminence of greater
(lateral) tuberosity (+).

Distal humerus (figs. 39-41):

I - Greatest breadth of distal condyle.
Measured parallel to longitudinal axis of
condyle on anterior surface (+).

J - Length (medial) of trochlea from bottom of
shallow depression immediately caudal to
proximal edge of articular surface of
trochlea to bottom of shallow depression
immediately caudal to distal edge of
articular surface (+).

K - Greatest diagonal depth of medial
epicondyle from bottom of shallow
depression immediately caudal to proximal
edge of articular surface of trochlea
(same as in J) (+).

L - Minimum depth of shaft from proximal end
of radial fossa to proximal end of
olecranon fossa. Measured on medial side
of shaft with tip of calipers in fossa (-).

M - Breadth of distal condyle at proximal edge of articular surface (+).

N - Length of capitulum. Measured perpendicular to distal edge of lateral epicondyle (+).

O - Length of lateral epicondyle from (and perpendicular to) distal edge of lateral epicondyle to proximal eminence of lateral epicondyloid crest (+).

Proximal radius (figs. 42-43):

A - Greatest breadth of proximal end (+).

B - Greatest depth of proximal end (+).

C - Depth of capitular articular surface. Measured from caudal edge of capitular articular surface at eminence formed by lateral edge of lateral ulnar facet (-).

D - Breadth between lateral and medial ulnar facets. Measured on caudal edge of capitular and trochlear articular surfaces between eminences formed by lateral edge of lateral ulnar facet (same as in C) and medial edge of medial ulnar facet (-).

E - Depth of sagittal ridge. Measured from eminence at medial edge of lateral ulnar facet on caudal edge of proximal articular surface (-).

F - Length of proximal ulnar articulation from

proximal end of interosseous space to
proximal eminence formed by medial edge of
medial ulnar facet (same as in D).

Measured on caudal surface of radius (-).

Distal radius (figs. 44-46):

G - Greatest breadth of distal end between
points of lateral and medial epiphyseal
fusion (+).

H - Length (medial) of distal epiphysis from
distal extremity of carpal articular
surface to proximal edge of prominent
tuberosity for attachment of medial carpal
ligament (point at or slightly proximal to
line of epiphyseal fusion; same as in
G) (+).

I - Greatest breadth of articular surfaces or
facets of intermediate and radial carpals.
Measured diagonally from lateral edge of
intermediate carpal facet to prominent
tuberosity for attachment of medial
collateral ligament of carpal joint (point
on medial face of distal trochlea, antero-
distal to tuberosity used in G and H) (+).

J - Minimum breadth of radial carpal facet.
Measured parallel to I at caudal edge of
facet (+).

K - Greatest breadth of radial carpal facet.
Measured parallel to I and J from lateral
edge of facet (same as in J) to prominent
tuberosity for attachment of medial
collateral ligament of carpal joint (same
as in I) (+).

Proximal femur (figs. 47-49):

A - Greatest breadth of proximal end. Measured
perpendicular to proximolateral surface of
greater trochanter (+).

B - Breadth of head and neck from lateral
margin of fovea capitis to most lateral
point of articular surface on proximal side
of neck (edge of fusion line). Point lies
caudal to longitudinal axis of neck (+).

C - Greatest craniocaudal diameter of head.
Identical to DC of von den Driesch (1976:
84-85) (+).

D - Greatest craniocaudal depth of greater
trochanter (+).

E - Minimum craniocaudal depth of neck.
Measured perpendicular to longitudinal axis
of neck just lateral to line of fusion (+).

F - Greatest craniocaudal depth of shaft at
lesser trochanter. Measured perpendicular
to line passing through fovea capitis and

center of greater trochanter (-).

G - Minimum length from most distal point on
lateroanterior margin of greater
trochanter to proximal saddle between
greater trochanter and neck (-).

Distal femur (figs. 50-53):

H - Greatest breadth of medial condyle (-).

I - Greatest breadth of distal end. Determined
with measuring box (+).

J - Greatest length of lateral condyle from
bottom of shallow depression immediately
cranial to proximal edge of articular
surface to distal edge of articular
surface (-).

K - Greatest depth (medial) of distal end from
cranial surface of medial ridge of trochlea
to caudal surface of medial condyle (+).

L - Greatest length of medial condyle from
bottom of shallow depression immediately
cranial to proximal edge of articular
surface to distal edge of articular
surface (-).

M - Greatest breadth of lateral condyle (-).

N - Greatest breadth of trochlea. Measured
perpendicular to lateral ridge (-).

0 - Minimum depth from cranial surface of

medial ridge of trochlea (same as in K) to
nearest point on caudal surface of shaft.
Measured on medial side of specimen (-).

Proximal tibia (fig. 54):

A - Greatest breadth of proximal end.
    Determined with measuring box (+).

B - Greatest depth from posteromedial
    extremity of medial condyle to anterior
    extremity of cranial border (-).

C - Breadth from posteromedial extremity of
    medial condyle (same as in B) to interior
    of extensor sulcus (+).

D - Greatest depth from posterolateral
    extremity of lateral condyle to anterior
    extremity of cranial border (same as in
    B) (-).

E - Breadth of medial condyle. Measured from
    posteromedial extremity of medial condyle
    (same as in B) to posterolateral edge of
    medial intercondylar tubercle (oblique to
    horizontal plane of proximal epiphysis)
    (-).

F - Breadth of lateral condyle. Measured from
    lateral extremity of lateral condyle to
    posteromedial edge of lateral intercon-
    dylar tubercle (oblique to horizontal

<u>Distal</u> <u>metapodial</u> (figs. 59-60):

    D - Greatest breadth of distal end. Measured
        at line of epiphyseal fusion (+).

    E - Breadth of medial condyle (Payne 1969:
        296) (+).

    F - Breadth of lateral condyle (Payne 1969:
        296) (+).

    G - Depth of medial trochlea (Payne 1969:
        296) (-).

    H - Depth of lateral trochlea (Payne 1969:
        296) (-).

    I - Depth of medial sagittal ridge (+).

    J - Depth of lateral sagittal ridge (+).

RESULTS

It must be stressed that the criteria developed in the present study are designed to augment, not replace, existing techniques for sexing bison limb elements. More important, the present criteria should be applied to fragmentary speci- mens only if the specimens are known to be mature (i.e., fused proximal humerus fragments, but not fused distal hum- erus fragments, etc.). The effectiveness of the various criteria developed here is indicated in tables 31 and 32 below; these tables summarize the results of a large series of bivariate crossplots. If the sexes were clearly and unambiguously separated in the plot for a given pair of plane of proximal epiphysis) (-).

G - Depth from posterior edge of depression
in center of intercondylar eminence to
anterior extremity of cranial border
(oblique to horizontal plane of proximal
epiphysis) (-).

Distal tibia (figs. 55-56):

H - Greatest breadth of distal end (+).

I - Greatest depth of distal end (+).

J - Greatest breadth of lateral and medial
articular grooves (-).

K - Length of medial malleolus from most dista
extremity to proximal margin of tuberosity
or roughness immediately proximal to media
malleolus (-).

L - Greatest diagonal distance from proximal
margin of tuberosity or roughness
immediately proximal to medial malleolus
(same as in K) to most distal extremity of
sagittal prominence or spine (-).

Proximal metapodial (figs. 57-58):

A - Greatest breadth of proximal end (+).

B - Greatest depth of proximal end (+).

C - Greatest breadth of articular facet of
fused 2d-3d carpal (metacarpal), or of
fused 2d-3d tarsal (metatarsal) (-).

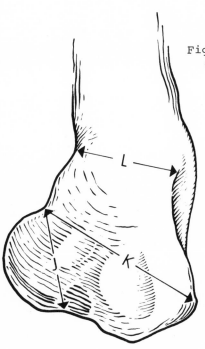

Fig. 40. Distal humerus
(right), medial view.

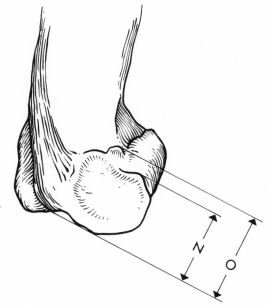

Fig. 41. Distal humerus
(right), lateral view.

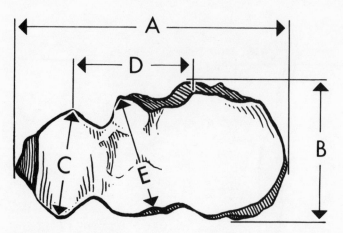

Fig. 42. Proximal radius (right), proximal view.

Fig. 43. Proximal radius (right), caudal view (position of ulna shown by dashed line).

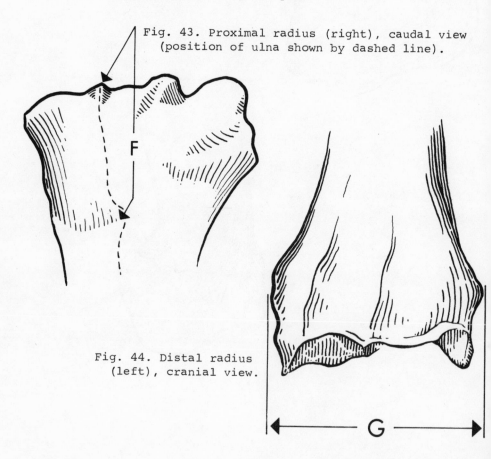

Fig. 44. Distal radius (left), cranial view.

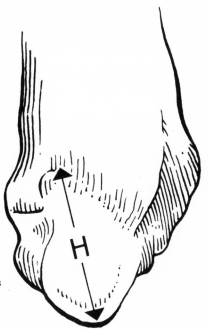

Fig. 45. Distal radius
(left), medial view.

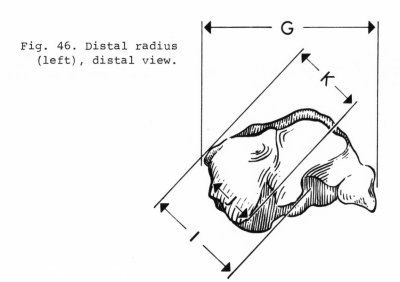

Fig. 46. Distal radius
(left), distal view.

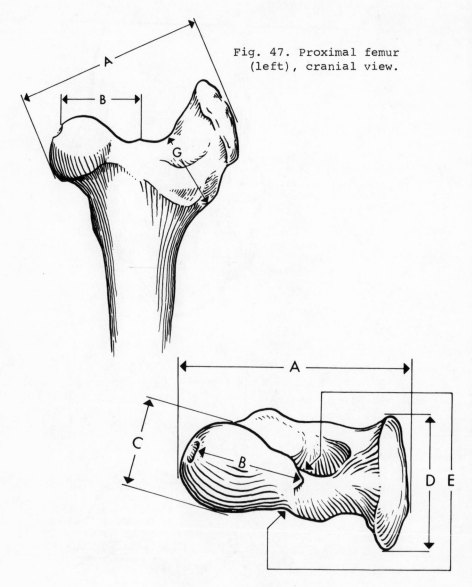

Fig. 47. Proximal femur
(left), cranial view.

Fig. 48. Proximal femur (left), proximal view.

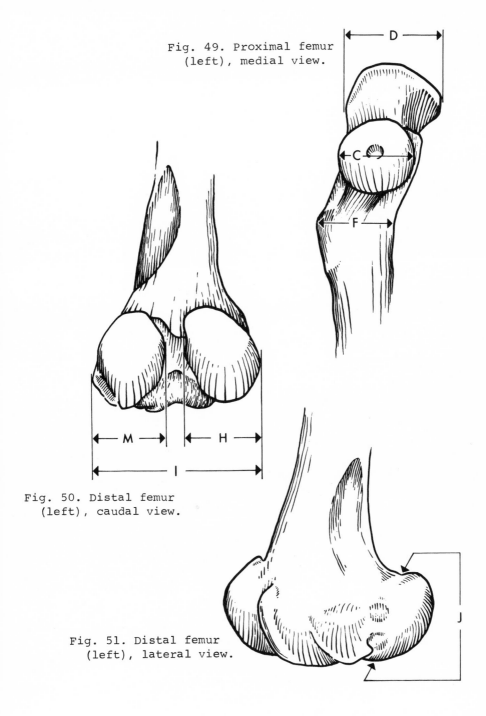

Fig. 49. Proximal femur
(left), medial view.

Fig. 50. Distal femur
(left), caudal view.

Fig. 51. Distal femur
(left), lateral view.

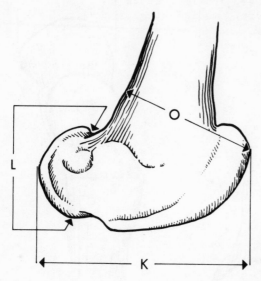

Fig. 52. Distal femur
(left), medial view.

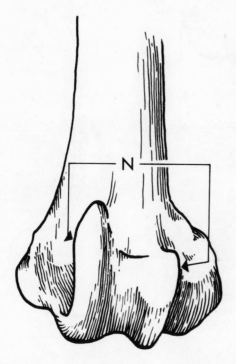

Fig. 53. Distal femur
(left), cranial view.

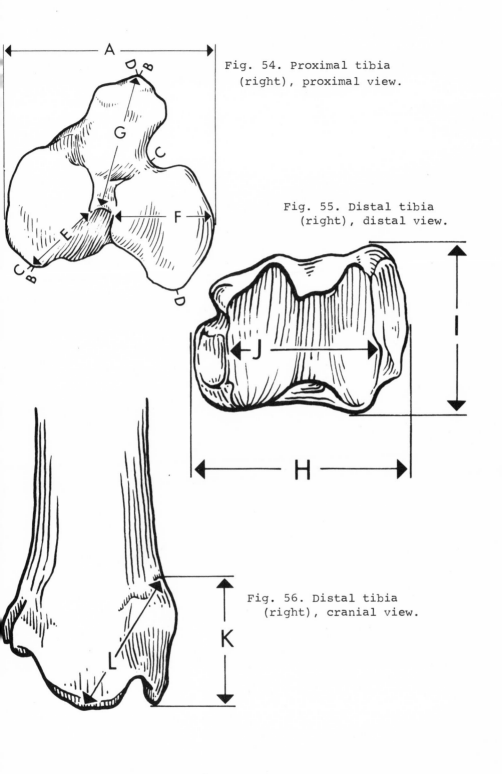

Fig. 54. Proximal tibia (right), proximal view.

Fig. 55. Distal tibia (right), distal view.

Fig. 56. Distal tibia (right), cranial view.

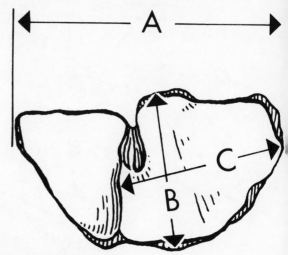

Fig. 57. Proximal metacarpal (right), proximal view.

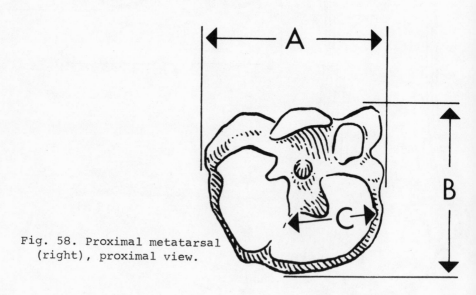

Fig. 58. Proximal metatarsal
(right), proximal view.

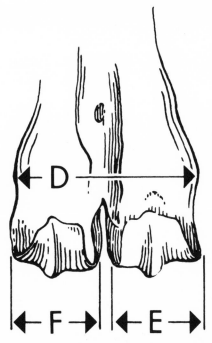

Fig. 59. Distal metapodial (right), cranial view.

Fig. 60. Distal metapodial (right), distal view.

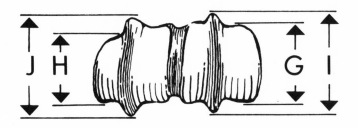

Table 19

Measurements of bison proximal humerus

Catalog Number[1]	Sex[2]	Side[3]	A	B	C	D	E	F	G	H
N534060[4]	1	2	11.42	8.60	13.00	11.50	9.34	13.33	10.06	12.04
N172689	1	1	12.70	9.73	13.50	13.20	10.15	15.50	10.83	14.19
N249894	1	1	12.10	9.74	13.00	12.30	8.78	13.62	10.22	12.50
NO16297	1	1	11.47	9.08	13.20	12.00	8.98	14.09	10.31	12.51
N200331	1	2	11.60	----	12.50	11.50	8.79	13.77	10.65	12.44
NO00839	1	2	11.90	9.23	12.90	12.50	9.13	14.12	10.57	12.29
NO49630	1	2	12.23	9.17	13.30	12.10	8.51	13.74	10.58	12.85
N251147	2	2	9.00	6.71	10.80	9.50	7.81	11.60	7.93	9.58
N251147[5]	2	2	10.85	6.69	10.55	9.50	7.96	11.59	8.04	9.49
NO63363	1	2	10.85	8.20	11.40	11.60	7.90	13.24	10.24	11.69
NO49571	1	2	11.50	8.78	12.00	12.30	8.19	13.65	10.82	11.96
N250719	2	2	9.15	7.90	10.80	10.50	7.43	11.84	7.68	10.30
N250719	2	1	9.35	7.63	10.80	10.50	7.54	11.94	7.93	10.59
N286873	1	2	11.50	8.82	12.70	11.80	8.69	14.04	10.43	12.36
NO22374	1	2	10.85	9.03	12.71	12.10	9.15	13.38	9.52	12.50
NO22375	1	2	10.15	8.15	11.50	10.80	8.90	13.00	9.01	11.33
NO22377	1	2	12.45	10.15	13.50	12.60	9.90	14.46	10.51	13.40
NO22663	1	2	10.83	8.25	12.30	11.40	8.81	13.34	9.08	11.97
NO49760	2	1	9.68	7.44	10.70	9.80	6.97	11.50	8.17	10.30
NO49760	2	2	9.40	7.39	10.50	9.80	6.90	11.33	8.00	10.09
N175783	1	2	11.65	9.36	12.60	12.30	8.38	13.26	10.38	12.10
N197705	2	2	9.62	7.68	10.90	10.00	7.53	11.81	7.94	9.97
N197705	2	1	9.40	7.46	11.10	10.10	7.57	12.02	7.73	9.99
NO22669	2	2	9.18	7.43	10.50	9.60	7.80	11.94	7.72	10.21
NO22669	2	1	9.03	7.23	10.60	9.80	7.76	11.86	7.49	10.17
NO49556	2	2	9.16	7.12	10.10	9.60	7.10	11.32	7.63	10.04
NO49631	1	2	11.16	9.01	11.70	11.60	8.57	13.93	10.28	12.29
NO22666	1	2	10.18	8.45	11.90	11.50	8.85	12.69	8.56	11.36
NO22668[4]	1	2	10.47	8.00	11.50	10.80	8.53	12.66	9.56	11.30
N176194	1	2	10.98	8.41	12.70	11.70	9.35	13.80	9.79	11.88
HO00092	2	2	9.55	7.34	10.90	9.90	7.67	11.89	8.06	10.40
HO20995	2	2	9.36	6.81	10.50	9.40	7.26	11.43	7.98	10.30
M114227	2	2	9.22	7.15	10.60	9.50	7.63	11.82	8.01	10.14

[1]H (Harvard University, Museum of Comparative Zoology); M (University of Michigan, Museum of Zoology); N (National Museum of Natural History).
[2]Male (1): female (2).
[3]Left (1): right (2).
[4]Proximal epiphysis fusing.
[5]Specimen slightly arthritic.

Table 20

Measurements of bison distal humerus

Catalog Number[1]	Sex[2]	Side[3]	I	J	K	L	M	N	O
N534060[4]	1	2	9.20	5.91	9.66	5.43	8.66	4.21	---
N172689	1	2	10.50	6.50	10.22	5.57	9.72	4.63	---
N172689	1	1	10.80	6.45	10.68	6.05	9.72	4.69	---
N249894	1	1	9.80	6.10	10.23	5.67	9.04	3.99	---
N016297	1	1	9.80	6.43	9.95	6.10	9.22	4.17	5.34
N200331	1	2	9.70	6.41	10.03	5.90	9.26	4.50	5.66
N000839	1	2	8.60	6.59	10.00	6.43	9.33	4.28	5.93
N049630	1	2	8.60	5.58	9.43	5.39	8.85	4.00	5.20
N251147	2	1	7.20	5.11	7.76	4.71	7.19	3.53	3.92
N251147[5]	2	2	7.20	5.17	7.93	4.70	7.39	3.47	3.97
N063363	1	2	8.30	5.61	8.92	5.58	8.33	4.09	5.22
N049571	1	2	9.10	5.90	9.31	5.86	8.82	4.10	5.92
N250719	2	2	7.50	4.90	8.06	4.33	7.24	3.34	3.94
N250719	2	1	7.60	5.00	8.19	4.23	7.23	3.30	3.85
N286873	1	2	9.30	5.77	9.77	5.81	9.19	4.31	5.49
N022374	1	2	9.20	6.15	9.42	5.68	9.08	4.29	5.80
N022375	1	2	8.80	5.71	8.78	5.38	8.77	4.03	4.56
N022377	1	2	9.40	6.63	9.99	6.01	9.35	4.50	5.54
N022663	1	2	8.40	5.94	9.03	5.34	8.60	4.07	5.12
N049760	2	1	7.40	4.91	7.87	4.44	7.56	3.26	3.92
N049760	2	2	7.50	4.82	7.71	4.43	7.50	3.36	3.90
N175783	1	2	8.50	5.80	9.35	6.00	8.25	3.74	5.70
N197705	2	2	7.90	5.13	8.53	4.71	7.68	3.49	4.15
N197705	2	1	7.90	5.14	8.77	4.87	7.72	3.42	4.31
N022669	2	2	7.80	5.23	8.32	4.80	7.97	3.70	4.58
N022669	2	1	7.90	5.15	8.35	4.84	7.98	3.65	4.60
N049556	2	2	8.00	4.99	7.47	4.44	7.17	3.37	3.87
N049631	1	2	8.60	5.93	9.33	5.73	9.00	4.05	5.22
N022666	1	2	8.90	5.90	8.77	5.31	8.36	4.00	4.81
N022668[4]	1	2	9.10	5.54	9.10	5.29	8.93	4.14	4.96
N176194	1	2	8.80	5.78	9.66	5.67	8.63	3.84	4.77
H000092	2	2	7.80	5.18	8.34	4.62	7.54	3.78	4.02
H020095	2	2	7.50	5.14	7.82	4.63	7.25	3.45	3.53
M114227	2	2	8.00	4.92	7.59	4.33	7.50	3.57	4.29

[1]H (Harvard University, Museum of Comparative Zoology); M (University of Michigan, Museum of Zoology); N (National Museum of Natural History).
[2]Male (1); female (2).
[3]Left (1); right (2).
[4]Proximal epiphysis fusing.
[5]Specimen slightly arthritic.

Table 21

Measurements of bison proximal radius

Catalog Number[1]	Sex[2]	Side[3]	A	B	C	D	E	F
N534060[4]	1	2	10.21	5.32	3.53	5.69	4.01	6.22
N172689	1	2	11.49	5.80	4.09	5.91	4.43	6.91
N249894	1	2	10.32	5.45	3.46	5.35	3.87	6.01
N016297	1	2	10.66	5.32	3.67	5.48	4.37	6.39
N200331	1	2	10.27	5.45	3.68	5.69	4.18	6.13
N000839	1	2	10.48	5.41	3.70	6.07	4.43	6.43
N049630	1	2	10.12	5.20	3.43	5.62	3.91	6.29
N251147	2	2	8.38	4.46	2.69	4.79	3.40	5.70
N063363	1	2	9.43	5.11	3.48	5.26	3.94	6.09
N049571	1	2	10.00	5.18	3.60	5.41	3.78	6.57
N250719	2	2	8.37	4.48	2.74	4.98	3.48	5.31
N286873	1	2	10.55	5.18	3.49	5.78	4.09	6.67
N022374	1	2	10.28	5.28	3.68	5.83	4.09	6.47
N022375	1	2	9.60	5.15	3.58	5.49	3.90	6.21
N022377	1	2	10.37	5.67	3.57	6.38	4.30	6.71
N022663	1	2	9.56	5.14	3.54	5.41	4.09	6.70
N049631	1	2	10.17	5.56	3.55	5.35	4.09	6.21
N049760	2	2	8.76	4.44	2.90	4.85	3.30	5.50
N175783	1	2	10.08	4.77	3.27	5.79	3.77	6.41
N197705	2	2	8.75	4.57	2.99	4.97	3.68	5.67
N022669	2	2	8.85	4.63	3.00	4.92	3.75	5.87
N049556	2	1	8.19	4.47	2.85	4.81	3.19	5.32
N022666	1	2	9.34	5.03	3.28	5.25	3.83	5.91
N176194	1	2	9.87	5.36	3.31	5.60	4.20	6.60
H000092	2	2	8.47	4.51	2.79	4.67	3.69	5.67
H020995	2	2	8.38	4.53	2.91	4.50	3.54	6.05
M114227	2	2	8.92	4.50	2.82	4.62	3.49	5.17

[1] H (Harvard University, Museum of Comparative Zoology); M (University of Michigan, Museum of Zoology); N (National Museum of Natural History).

[2] Male (1); female (2).

[3] Left (1); right (2).

[4] Distal epiphysis fusing.

Table 22

Measurements of bison distal radius

Catalog Number[1]	Sex[2]	Side[3]	G	H	I	J	K
N534060[4]	1	2	9.12	5.14	5.18	2.07	3.57
N172689	1	2	10.22	6.53	5.88	2.42	4.35
N249894	1	2	9.35	4.77	5.31	2.36	3.79
NO16297	1	2	9.18	5.25	5.15	2.18	3.62
N200331	1	2	9.35	4.78	5.20	2.09	3.65
NO00839	1	2	9.47	5.38	5.72	2.26	3.96
NO49630	1	2	9.19	5.64	5.35	2.08	3.89
N251147	2	2	7.51	4.23	4.27	1.54	2.91
NO63363	1	2	8.86	5.13	5.22	2.03	3.81
NO49571	1	2	9.38	4.99	5.18	1.90	3.85
N250719	2	2	7.50	4.16	4.38	1.54	2.95
N286873	1	2	9.89	4.86	5.53	2.23	3.81
NO22374	1	2	9.70	4.77	5.16	1.97	3.52
NO22375	1	2	8.53	4.70	5.24	1.93	3.60
NO22377	1	2	9.53	5.13	5.62	2.14	3.94
NO22663	1	2	8.35	4.63	4.74	1.82	3.42
NO49631	1	2	9.28	4.96	5.50	2.02	3.93
NO49760	2	2	7.58	4.24	4.42	1.75	3.03
N175783	1	2	9.20	4.70	5.12	1.71	3.68
N197705	2	2	7.83	4.48	4.49	1.45	3.06
NO22669	2	2	8.40	4.39	4.57	1.66	2.99
NO49556	2	1	7.43	4.07	3.79	1.36	2.69
NO22666	1	2	8.51	4.87	4.92	1.69	3.34
N176194	1	2	9.46	4.98	5.07	2.30	3.44
HO00092	2	2	7.29	4.37	4.39	1.63	2.92
HO20995	2	2	7.45	4.04	4.18	1.56	2.91
M114227	2	2	7.56	4.07	4.32	1.50	2.92

[1]H (Harvard University, Museum of Comparative Zoology); M
(University of Michigan, Museum of Zoology); N (National
Museum of Natural History).
[2]Male (1); female (2).
[3]Left (1); right (2).
[4]Distal epiphysis fusing.

Table 23

Measurements of bison proximal metacarpal

Catalog Number[1]	Sex[2]	Side[3]	A	B	C
N049631	1	2	7.72	4.55	4.23
N049760	2	2	6.45	3.75	3.67
N175783	1	2	7.88	4.29	4.13
N197705	2	2	6.65	4.05	3.61
N022669	2	2	6.86	4.13	3.85
N049556	2	1	6.06	3.40	3.42
N154570	1	2	7.29	4.36	4.14
N176194	1	2	7.48	4.64	3.91
N022663	1	2	7.28	4.40	4.00
N022664	1	2	7.08	4.26	4.13
N022377	1	2	7.84	4.64	4.53
N286873	1	2	7.87	4.78	4.47
N022375	1	2	7.20	4.18	4.04
N063363	1	2	7.42	4.12	3.98
N049571	1	2	7.83	4.62	4.23
N250719	2	2	6.38	3.60	3.75
N251147	2	2	6.31	3.43	3.61
N000839	1	2	7.94	4.55	4.30
H000092	2	1	6.34	4.04	3.54
H020995	2	2	6.53	4.17	3.63
M114227	2	2	6.14	3.93	3.62

[1]H (Harvard University, Museum of Comparative Zoology); M (University of Michigan, Museum of Zoology); N (National Museum of Natural History).
[2]Male (1); female (2).
[3]Left (1); right (2).

Table 24

Measurements of bison distal metacarpal

Catalog Number[1]	Sex[2]	Side[3]	D	E	F	G	H	I	J
N049631	1	2	7.59	3.65	3.42	3.05	2.73	4.05	3.93
N049760	2	2	6.39	3.23	3.04	2.71	2.46	3.51	3.39
N175783	1	2	7.69	3.37	3.34	2.87	2.63	3.88	3.76
N197705	2	2	6.53	3.11	2.96	2.80	2.56	3.63	3.57
N022669	2	2	6.55	3.21	3.04	2.84	2.56	3.73	3.59
N049956	2	1	6.03	2.86	2.65	2.46	2.27	3.35	3.20
N154570	1	2	6.81	3.44	3.28	3.05	2.69	3.96	3.76
N176194	1	2	7.22	3.59	3.37	3.17	2.81	3.91	3.82
N022663	1	2	7.55	3.76	3.52	3.28	2.77	4.10	3.74
N022664	1	2	6.97	3.69	3.38	3.14	2.71	4.03	3.89
N022377	1	2	7.42	3.92	3.67	3.51	3.02	4.53	4.23
N286873	1	2	8.00	3.92	3.68	3.30	2.76	4.24	3.99
N022375	1	2	6.84	3.45	3.20	2.97	2.73	3.91	3.69
N063363	1	2	7.28	3.41	3.34	2.91	2.61	3.81	3.65
N049571	1	2	7.74	3.55	3.50	3.14	2.86	4.16	4.00
N250719[4]	2	2	6.55	3.15	2.91	2.75	2.45	3.55	3.40
N251147	2	2	6.31	2.97	2.83	2.61	2.39	3.40	3.31
N000839	1	1	8.00	3.94	3.74	3.19	2.84	4.15	3.91
H000092	2	2	6.22	3.18	2.94	2.68	2.39	3.53	3.43
H020995	2	2	6.28	---	---	---	---	---	---
M114227	2	2	6.42	3.11	2.87	2.63	2.36	3.48	3.37

[1] H (Harvard University, Museum of Comparative Zoology); M (University of Michigan, Museum of Zoology); N (National Museum of Natural History).
[2] Male (1); female (2).
[3] Left (1); right (2).
[4] Specimen slightly arthritic.

Table 25

Measurements of bison proximal femur

Catalog Number[1]	Sex[2]	Side[3]	A	B	C	D	E	F	G
N534060	1	1	-----	7.34	5.89	----	3.20	6.03	6.53
N534060	1	2	-----	7.64	5.85	----	3.17	5.95	6.55
N251147	2	1	12.20	6.26	5.26	6.78	2.78	5.60	5.34
N251147	2	2	12.35	6.22	5.29	6.98	2.69	5.48	5.48
N172689	1	1	16.00	8.42	6.60	----	3.22	7.18	6.93
N249894	1	1	16.03	7.45	6.16	----	2.98	6.07	6.43
N249894	1	2	14.53	7.13	5.87	9.20	3.20	6.25	8.21
NO16297	1	2	15.22	7.86	6.10	9.94	3.07	6.57	6.36
N200331	1	1	14.54	6.97	5.67	----	2.92	5.93	6.41
N200331	1	2	14.42	7.05	5.65	8.93	2.91	5.79	6.39
NO00839	1	2	14.87	7.50	6.15	8.59	3.05	5.95	6.10
NO00839	1	1	14.78	7.78	6.14	8.50	3.30	6.04	6.27
NO00839	1	1	14.47	7.07	5.75	8.89	3.63	6.43	6:72
NO49630	1	2	13.93	6.90	5.97	8.84	3.29	6.23	5.83
NO49630	1	1	14.08	6.85	5.47	8.65	2.97	6.32	5.86
N269169	1	1	14.13	7.11	5.82	8.74	3.12	6.23	6.23
N269169	1	2	14.14	7.15	5.87	8.53	3.11	6.00	6.23
NO63363	1	1	13.66	6.72	5.39	7.92	2.64	6.14	5.40
NO63363	1	2	13.60	6.78	5.40	7.92	2.70	6.12	5.93
NO49571	1	2	14.98	7.45	5.81	8.88	3.20	6.24	6.34
N250719	2	1	12.50	6.21	5.18	7.80	2.66	5.29	5.28
N250719	2	2	12.40	6.34	5.17	7.50	2.74	----	4.99
N286873	1	1	14.99	7.10	5.69	8.90	2.92	6.72	5.99
NO22374	1	2	14.76	7.07	5.67	9.24	3.17	6.48	6.17
NO22375	1	1	13.82	7.12	5.44	8.30	2.50	5.31	5.76
NO22377	1	2	14.44	7.15	5.65	8.68	2.79	6.04	6.02
NO22663	1	2	14.23	7.17	5.82	8.81	3.08	6.34	6.29
NO49631	1	1	14.20	6.74	5.72	9.15	2.88	6.38	6.32
NO49760	2	1	11.99	6.04	4.98	7.62	2.67	5.40	5.36
NO49760	2	2	11.86	6.11	4.91	7.41	2.65	5.16	5.57
N175783	1	1	14.57	6.84	5.43	9.83	3.16	6.50	6.39
N197705	2	1	12.71	6.40	5.50	7.48	2.62	5.43	5.05
N197705	2	2	12.85	6.51	5.34	7.54	2.66	5.62	5.34
NO22669	2	1	12.50	6.25	5.48	6.80	2.74	5.78	5.39
NO22669	2	2	12.83	6.46	5.40	6.82	2.85	5.59	5.32
NO49556	2	2	12.26	6.02	5.02	7.72	2.51	5.06	4.92
NO49631	1	2	14.46	6.68	5.66	9.34	2.79	6.12	5.96
NO22666	1	2	13.34	6.60	5.26	8.35	2.86	5.68	5.84
NO22668	1	2	14.40	6.62	5.52	8.41	2.72	5.76	5.88
N176194	1	2	13.96	7.14	5.81	8.86	2.61	5.81	6.07
HO00092	2	2	12.75	6.30	5.27	7.52	2.82	5.48	5.45
HO20995	2	2	11.52	6.02	5.13	6.79	2.49	5.54	5.56
M114227	2	2	12.25	6.00	5.19	7.03	2.28	5.01	5.56

[1]H (Harvard University, Museum of Comparative Zoology); M (University of Michigan, Museum of Zoology); N (National Museum of Natural History).
[2]Male (1); female (2).
[3]Left (1); right (2).

Table 26

Measurements of bison distal femur

Catalog Number[1]	Sex[2]	Side[3]	H	I	J	K	L	M	N	O
N534060	1	1	5.25	11.98	7.40	15.25	6.10	5.41	6.68	10.69
N534060	1	2	5.15	12.00	7.00	---	6.05	5.10	---	---
N172689	1	1	6.10	13.09	7.58	15.90	6.77	6.09	6.92	11.21
N249894	1	2	5.09	11.19	7.26	14.74	5.75	5.88	6.37	10.78
N249894	1	2	5.26	11.83	6.82	15.50	5.60	6.03	6.21	11.03
N016297	1	2	5.06	12.04	7.54	15.23	6.21	5.87	6.50	10.95
N200331	1	1	---	---	7.50	---	5.82	---	---	---
N200331	1	2	5.00	12.07	7.17	15.26	5.75	5.88	6.14	10.64
N000839	1	2	5.31	11.80	7.00	15.21	6.02	6.75	6.60	10.81
N000839	1	1	5.12	11.79	6.98	14.69	6.05	6.77	6.48	10.83
N049630	1	2	0.00	11.60	6.58	14.78	5.80	6.13	6.22	10.28
N049630	1	1	4.95	11.66	6.97	14.63	5.79	6.10	6.19	10.25
N269169	1	1	5.18	12.74	7.20	14.72	6.00	5.84	6.05	10.77
N269169	1	2	5.30	12.35	7.06	14.70	5.86	5.86	6.22	10.33
N063363	1	1	4.82	11.12	6.35	13.98	5.56	5.44	6.06	9.40
N063363	1	2	4.84	11.20	6.23	13.97	5.50	5.34	6.40	9.40
N049571	1	2	4.87	11.81	7.00	14.77	6.01	5.76	6.87	10.47
N250719	2	1	4.31	10.05	5.78	12.99	4.92	4.74	5.40	9.29
N286873	1	1	5.16	12.28	7.01	14.73	5.90	6.05	5.94	10.68
N022374	2	1	4.93	12.03	6.69	14.96	6.00	5.74	6.39	10.48
N022375	1	1	4.81	11.13	6.32	13.83	5.40	5.30	5.86	9.86
N022377	2	1	4.94	11.44	6.91	14.84	5.91	6.14	6.55	10.58
N022663	1	1	5.05	11.91	6.80	14.85	6.24	6.09	6.50	10.54
N049631	1	1	5.45	11.99	6.60	14.88	5.68	5.52	6.38	10.63
N049760	2	1	4.13	9.79	5.39	13.31	4.93	4.59	5.49	9.14
N049760	2	2	4.07	9.70	5.49	13.35	4.83	4.48	5.35	9.15
N175783	1	1	4.81	11.39	6.10	14.73	5.49	5.44	5.95	10.42
N197705	2	1	4.27	10.03	5.88	13.68	5.29	4.37	5.73	9.69
N197705	2	2	4.27	10.30	6.08	13.50	5.15	4.54	5.69	9.72
N022669	2	1	4.52	10.60	6.08	13.68	4.92	5.00	5.33	9.71
N022669	2	2	4.37	10.64	6.02	13.72	5.04	4.50	5.29	9.70
N049556	2	2	4.02	9.70	5.34	12.78	4.84	4.19	5.12	9.12
N049631	1	2	5.38	11.78	6.41	14.83	5.80	5.12	6.69	10.45
N022666	1	2	4.81	11.21	6.29	14.20	5.69	5.20	5.62	10.14
N022668	2	2	4.82	11.35	6.23	14.17	5.54	5.22	6.11	10.04
N176194	1	2	5.09	11.55	6.60	14.48	6.09	5.64	6.50	10.43
H000092	2	2	4.14	10.24	5.71	13.18	5.03	4.52	5.35	9.33
H020995	2	2	4.08	10.22	5.46	12.44	5.01	4.51	4.87	8.48
M114227	2	2	4.47	10.06	5.88	13.21	5.16	4.02	5.24	9.22

[1]H (Harvard University, Museum of Comparative Zoology); M (University of Michigan, Museum of Zoology); N (National Museum of Natural History).
[2]Male (1); female (2).
[3]Left (1); right (2).

Table 27

Measurements of bison proximal tibia

Catalog Number[1]	Sex[2]	Side[3]	A	B	C	D	E	F	G
N534060[4]	1	2	11.85	11.94	9.42	11.80	6.24	5.90	9.49
N249894	1	2	12.25	12.45	9.27	12.19	6.07	6.53	9.22
N016297	1	2	12.32	12.38	9.55	12.78	6.32	6.48	9.74
N200331	1	2	12.43	12.01	9.40	12.28	6.13	6.67	9.61
N000839	1	2	12.22	12.20	9.88	11.78	6.52	6.66	9.74
N049630	1	2	12.26	11.82	9.24	11.88	6.29	6.12	9.23
N251147[5]	2	2	11.00	10.81	8.14	10.45	---	---	---
N063363	1	2	11.64	11.26	9.00	11.07	6.03	5.96	8.72
N049571	1	2	12.43	11.85	9.16	11.55	6.12	6.82	9.50
N250719	2	1	10.17	10.11	7.97	9.98	5.26	5.30	7.85
N286873	1	2	12.19	12.17	9.14	12.11	6.53	6.37	9.27
N022374	2	2	12.40	12.13	9.29	12.41	6.47	6.00	9.50
N022375	2	2	11.70	10.93	8.81	10.74	6.12	6.02	8.45
N022377	1	2	13.02	12.60	9.46	12.07	6.57	6.82	10.29
N022663	1	2	12.30	11.67	9.24	12.05	6.05	6.43	9.58
N049760	2	2	10.43	10.43	8.12	10.27	5.33	5.30	8.23
N175783	1	2	11.62	11.52	8.84	11.68	6.08	5.99	8.62
N197705	2	2	10.54	10.87	8.32	10.63	5.32	5.77	8.28
N022669	2	2	10.87	11.08	8.41	10.80	5.48	5.49	8.54
N049556	1	1	9.89	9.94	7.69	9.88	5.04	5.09	7.63
N049631	1	2	12.26	12.05	9.24	11.80	6.51	6.54	9.14
N022666	1	2	11.53	11.16	8.65	11.03	5.85	6.37	8.57
N022668[4]	1	2	11.87	11.46	9.01	11.39	5.98	6.16	8.98
N176194	1	2	11.88	12.06	9.19	12.12	6.33	6.13	9.59
H000092	2	2	10.72	10.43	8.12	10.42	5.33	5.95	8.22
H020995	2	1	10.30	10.07	7.93	10.09	4.93	5.31	7.74
M114227	2	2	10.68	10.59	8.05	10.64	5.33	5.58	8.21

[1] H (Harvard University, Museum of Comparative Zoology); M (University of Michigan, Museum of Zoology); N (National Museum of Natural History).
[2] Male (1); female (2).
[3] Left (1); right (2).
[4] Proximal epiphysis fusing.
[5] Specimen slightly arthritic.

Table 28

Measurements of bison distal tibia

Catalog Number[1]	Sex[2]	Side[3]	H	I	J	K	L
N534060[4]	1	2	7.75	5.55	5.50	4.38	5.65
N249894	1	2	7.92	5.73	5.33	4.50	5.59
N016297	1	2	7.76	5.59	5.56	5.02	5.95
N200331	1	2	7.91	5.34	5.23	5.18	5.86
N000839	1	2	7.56	5.84	5.25	5.10	6.01
N049630	1	2	7.45	5.55	5.10	4.87	5.85
N251147[5]	2	2	6.36	4.82	4.48	3.41	4.86
N063363	1	2	7.37	5.56	5.13	4.70	5.73
N049571	1	1	7.68	5.63	5.16	5.41	6.27
N250719	2	1	6.61	5.02	4.57	3.81	4.86
N286873	1	2	7.77	5.89	5.38	5.34	6.18
N022374	1	2	7.62	5.90	5.43	4.99	5.96
N022375	1	2	7.02	5.19	5.11	4.20	5.28
N022377	1	2	7.98	5.83	5.46	5.26	6.29
N022663	1	2	7.67	5.54	5.47	5.00	6.09
N049760	2	2	6.54	5.13	4.99	4.21	5.06
N175783	1	2	7.56	5.83	5.10	4.74	5.61
N197705	2	2	6.85	5.06	4.85	4.17	5.26
N022669	2	2	7.34	5.36	5.08	4.25	5.39
N049556	2	1	6.23	4.67	4.43	3.64	4.60
N049631	1	2	7.68	5.90	5.24	4.71	5.54
N022666	1	2	7.07	5.11	5.23	4.77	5.72
N022668[4]	1	2	7.36	5.31	5.44	4.86	5.70
N176194	1	2	7.31	5.47	5.38	4.88	5.75
H000092	2	2	6.75	4.94	4.77	3.91	4.99
H020995	2	1	6.55	5.04	4.63	3.90	4.98
M114227	2	2	6.73	4.81	4.69	3.95	5.03

[1] H (Harvard University, Museum of Comparative Zoology); M (University of Michigan, Museum of Zoology); N (National Museum of Natural History).
[2] Male (1); female (2).
[3] Left (1); right (2).
[4] Proximal epiphysis fusing.
[5] Specimen slightly arthritic.

Table 29

Measurements of bison proximal metatarsal

Catalog Number[1]	Sex[2]	Side[3]	A	B	C
N049631	1	2	6.06	5.47	3.07
N049760	2	2	5.04	4.77	2.69
N175783	1	2	6.68	6.10	3.18
N197705	1	2	5.11	5.07	2.73
N022669	2	2	5.16	5.18	2.65
N049556	2	2	4.68	4.63	2.45
N154570	1	2	5.23	5.31	2.61
N176194	1	2	5.56	5.33	2.73
N022664	1	2	5.45	5.50	2.95
N022377	1	2	6.28	5.90	3.02
N286873	1	2	5.98	5.73	3.05
N022375	1	2	5.57	5.01	2.59
N063363	1	2	6.27	5.26	2.92
N049571	1	2	6.12	5.69	3.04
N250719	2	2	5.03	4.71	2.56
N251147	2	2	4.85	4.69	2.40
N000839	1	2	5.85	5.55	2.78
H000092	2	2	4.99	4.93	2.46
H020995	2	2	4.85	4.95	2.58
M114227	2	1	5.30	4.87	2.45

[1]H (Harvard University, Museum of Comparative Zoology): M (University of Michigan, Museum of Zoology): N (National Museum of Natural History).
[2]Male (1); female (2).
[3]Left (1); right (2).

Table 30

Measurements of bison distal metatarsal

Catalog Number[1]	Sex[2]	Side[3]	D	E	F	G	H	I	J
NO49631	1	2	6.70	3.21	3.12	2.89	2.53	3.99	3.82
NO49760	2	2	5.91	2.83	2.70	2.62	2.47	3.56	3.48
N175783	1	2	6.84	2.90	2.82	2.80	2.50	3.87	3.71
N197705	2	2	6.08	2.74	2.58	2.76	2.46	3.66	3.51
NO22669	2	2	6.35	3.13	2.90	2.90	2.54	3.85	3.68
NO49556	2	2	5.43	2.46	2.39	2.39	2.19	3.33	3.21
N154570	1	2	6.21	3.02	2.87	2.70	2.49	3.81	3.61
N176194	1	2	6.44	3.08	2.92	2.82	2.58	3.75	3.64
NO22664	1	2	6.37	3.22	3.00	2.98	2.62	3.95	3.68
NO22377	1	2	6.81	3.36	3.18	3.23	2.91	4.28	4.17
N286873	1	2	6.86	3.20	3.06	2.92	2.64	4.15	3.92
NO22375	1	2	6.14	2.96	2.85	2.85	2.57	3.75	3.57
NO63363	1	2	6.60	3.02	2.80	2.80	2.39	3.67	3.46
NO49571	1	2	6.91	3.17	3.03	2.94	2.62	4.04	3.85
N250719	2	2	5.61	2.72	2.48	2.64	2.29	3.48	3.32
N251147	2	2	5.80	2.67	2.56	2.61	2.37	3.56	3.31
NO00839	1	2	6.88	3.32	3.18	3.08	2.80	4.10	3.88
HO00092	2	2	5.87	2.88	2.76	2.67	2.51	3.61	3.51
HO20995	2	2	5.70	---	---	---	---	---	---
M114227[4]	2	1	6.22	2.90	2.72	2.68	2.33	3.60	3.35

[1]H (Harvard University, Museum of Comparative Zoology); M (University of Michigan, Museum of Zoology); N (National Museum of Natural History).
[2]Male (1); female (2).
[3]Left (1); right (2).
[4]Specimen slightly arthritic.

Table 31

Effectiveness of criteria for sexing
front-limb elements of modern bison

PH C¹	PH R²	DH C	DH R	PR C	PR R	DR C	DR R	PMc C	PMc R	DMc C	DMc R
A:B	G	I:J	G	A:B	G	G:H	F	A:B	F	D:E	G
A:C	G	I:K	F	A:C	G	G:I	F	A:C	G	D:F	G
A:D	G	I:L	F	A:D	G	G:J	F	B:C	F	D:G	F
A:E	G	I:M	G	A:E	F	G:K	G	–	–	D:H	F
A:F	G	I:N	G	A:F	P	H:I	F	–	–	D:I	F
A:G	G	I:O	F	B:C	G	H:J	P	–	–	D:J	F
A:H	G	J:K	G	B:D	G	H:K	G	–	–	E:F	G
B:C	F	J:L	G	B:E	F	I:J	P	–	–	E:G	F
B:D	F	J:M	G	B:F	F	I:K	G	–	–	E:H	F
B:E	F	J:N	G	C:D	G	J:K	F	–	–	E:I	F
B:F	G	J:O	G	C:E	F	–	–	–	–	E:J	F
B:G	G	K:L	F	C:F	F	–	–	–	–	F:G	G
B:H	G	K:M	F	D:F	F	–	–	–	–	F:H	G
C:D	G	K:N	F	E:F	P	–	–	–	–	F:I	F
C:E	F	K:O	P	–	–	–	–	–	–	F:J	P
C:F	G	L:M	G	–	–	–	–	–	–	G:H	F
C:G	G	L:N	G	–	–	–	–	–	–	G:I	F
C:H	G	L:O	G	–	–	–	–	–	–	G:J	F
D:E	G	M:N	G	–	–	–	–	–	–	H:I	F
D:F	G	M:O	P	–	–	–	–	–	–	H:J	F
D:H	F	N:O	P	–	–	–	–	–	–	I:J	F
E:F	G	–	–	–	–	–	–	–	–	–	–
E:G	G	–	–	–	–	–	–	–	–	–	–
E:H	G	–	–	–	–	–	–	–	–	–	–
F:G	G	–	–	–	–	–	–	–	–	–	–
F:H	G	–	–	–	–	–	–	–	–	–	–
G:H	G	–	–	–	–	–	–	–	–	–	–

¹C (Criteria; colon denotes crossplot).
²R (Rating: G, good; F, fair; P, poor).

Table 32

Effectiveness of criteria for sexing
rear-limb elements of modern bison

PF		DF		PT		DT		PMt		DMt	
C¹	R²	C	R	C	R	C	R	C	R	C	R
A:B	G	H:I	G	A:B	G	H:I	F	A:B	P	D:E	P
A:C	G	H:J	G	A:C	G	H:J	F	A:C	P	D:F	P
A:D	G	H:K	G	A:D	G	H:K	F	B:C	P	D:G	P
A:E	F	H:L	G	A:E	G	H:L	G			D:H	P
A:F	F	H:M	G	A:F	G	I:J	P			D:I	P
A:G	G	H:N	G	B:C	P	I:K	P			D:J	P
B:C	P	H:O	F	B:D	P	I:L	P			E:F	P
B:D	G	I:J	G	B:E	G	J:K	P			E:G	P
B:E	P	I:K	G	B:F	P	J:L	P			E:H	P
B:F	F	I:L	G	B:G	P	K:L	P			E:I	P
B:G	F	I:M	G	C:D	G					E:J	P
C:D	P	I:N	G	C:E	G					F:G	P
C:E	P	I:O	F	C:F	P					F:H	P
C:F	P	J:K	P	C:G	G					F:I	P
C:G	P	J:L	P	D:E	P					F:J	P
D:E	P	J:M	P	D:F	G					G:H	P
D:F	P	J:N	P	D:G	P					G:I	P
D:G	P	J:O	P	E:F	G					G:J	P
E:F	P	K:L	P	E:G	G					H:I	P
E:G	P	K:M	P	F:G	P					H:J	P
F:G	P	K:N	P							I:J	F
		K:O	F								
		L:M	P								
		L:N	P								
		L:O	P								
		M:N	F								
		M:O	P								
		N:O	P								

¹C (Criteria; colon denotes crossplot).
²R (Rating: G, good; F, fair; P, poor).

References

Allen, C. E., D. C. Beitz, D. A. Cramer, and R. G. Kauffman. 1976. Biology of fat in meat animals. North Central Regional Research Publication 234. Madison: University of Wisconsin, College of Agricultural and Life Sciences.

Allen, Durward L. 1979. Wolves of Minong. Boston: Houghton Mifflin.

Anderson, A. E., D. E. Medin, and D. C. Bowden. 1972. Indices of carcass fat in a Colorado mule deer population. Journal of Wildlife Management 36(2):579-94.

Arasu, P., R. A. Field, W. G. Kruggel, and G. J. Miller. 1981. Nucleic acid content of bovine bone marrow, muscle and mechanically deboned beef. Journal of Food Science 46(4):1114-16.

Arthur, George W. 1975. An introduction to the ecology of early historic communal bison hunting among the Northern Plains Indians. Mercury Series, Archaeological Survey of Canada Paper 37. Ottawa: National Museum of Man.

Baerreis, David A., and Reid A. Bryson. 1966. Dating the Panhandle Aspect cultures. Oklahoma Anthropological Society Bulletin 14:105-16.

Baker, G., L. H. P. Jones, and I. D. Wardrop. 1959. Cause of wear in sheeps' teeth. Nature 184:1583-84.

Baker, Maurice F., and Francis X. Lueth. 1967. Mandibular cavity tissue as a possible indicator of condition in deer. Southeastern Association of Game and Fish Commissioners, Annual Conference Proceedings 20:69-74.

Barnes, Richard H. 1976. Energy. In Present knowledge in nutrition, 4th ed., ed. D. M. Hegsted et al., pp. 10-16. New York: Nutrition Foundation.

Bedord, Jean N. 1974. Morphological variation in bison metacarpals and metatarsals. In The Casper Site, ed. George C. Frison, pp. 199-240. New York: Academic Press.

_____. 1978. A technique for sex determination of mature bison metapodials. Plains Anthropologist Memoir 14:40-43.

Behrensmeyer, Anna K. 1975. The taphonomy and paleoecology of Plio-Pleistocene vertebrate assemblages east of Lake Rudolf, Kenya. Harvard University, Museum of Comparative Zoology Bulletin 146(10):473-578.

Bell, Robert E. 1958. A guide to the identification of certain American Indian projectile points. Special Bulletin 1. Oklahoma City: Oklahoma Anthropological Society.

Bender, Margaret M. 1968. Mass spectrometric studies of carbon 13 variations in corn and other grasses. Radiocarbon 10(2):468-72.

Bigwood, E. J., ed. 1972. Protein and amino acid functions. Oxford: Pergamon Press.

Binford, Lewis R. 1978. Nunamiut ethnoarchaeology. New York: Academic Press.

_____. 1981. Bones: Ancient men and modern myths. New York: Academic Press.

Binford, Lewis R., and Jack B. Bertram. 1977. Bone frequencies--and attritional processes. In For theory building in archaeology, ed. Lewis R. Binford, pp. 77-153. New York: Academic Press.

Boggino, Eloy J. A. 1970. Chemical composition and in vitro dry matter digestibility of range grasses. M.S. thesis, Animal Science, New Mexico State University.

Borchert, John R. 1950. The climate of the central North American grassland. Annals of the Association of American Geographers 40(1):1-39.

Brain, C. K. 1967. Hottentot food remains and their meaning in the interpretation of fossil bone assemblages. Namib Desert Research Station Scientific Papers 32:1-11.

_____. 1980. Some criteria for the recognition of bone-collecting agencies in African caves. In Fossils in the making, ed. Anna K. Behrensmeyer and Andrew P. Hill, pp. 107-30. Chicago: University of Chicago Press.

_____. 1981. The hunters or the hunted? An introduction to African cave taphonomy. Chicago: University of Chicago Press.

Brooks, P. M. 1978. Relationship between body condition and age, growth, reproduction and social status in impala, and its application to management. South African Journal of Wildlife Research 8(4):151-57.

Brooks, P. M., J. Hanks, and J. V. Ludbrook. 1977. Bone marrow as an index of condition in African ungulates. South African Journal of Wildlife Research 7(2):61-66.

Brown, Christopher L., and Carl E. Gustafson. 1979. A key to postcranial skeletal remains of cattle/bison, elk, and horse. Reports of Investigations 57. Pullman: Washington State University Laboratory of Anthropology.

Bryson, R. A., D. A. Baerreis, and W. M. Wendland. 1970. The character of late glacial and post-glacial climatic changes. In Pleistocene and Recent environments of the central Great Plains, ed. Wakefield Dort, Jr., and J. Knox Jones, Jr., pp. 53-74. Lawrence: University Press of Kansas.

Buffington, Lee C., and Carlton H. Herbel. 1965. Vegetational changes on a semidesert grassland range from 1858 to 1963. Ecological Monographs 35(2):139-64.

Bureau of Land Management. 1979. East Roswell grazing environmental statement. United States Department of the Interior, Bureau of Land Management, Roswell District, New Mexico.

Castetter, Edward F. 1956. The vegetation of New Mexico. Annual Research Lecture 3. Albuquerque: University of New Mexico.

Caswell, H., F. Reed, S. N. Stephenson, and P. A. Werner. 1973. Photosynthetic pathways and selective herbivory: A hypothesis. American Naturalist 107(956):465-80.

Chaney, Margaret S., and Margaret L. Ross. 1971. Nutrition. 8th ed. Boston: Houghton Mifflin.

Collins, Michael B. 1966. The Andrews Lake Sites: Evidence of semi-sedentary prehistoric occupation in Andrews County, Texas. Midland Archaeological Society Bulletin 1:27-43.

_____. 1971. A review of Llano Estacado archeology and eth-nohistory. Plains Anthropologist 16:85-104.

Colorado State University Extension Service. 1981. Handbook of Colorado native grasses. Bulletin 450-A. Fort Collins: Colorado State University Extension Service.

Cook, C. W., R. D. Child, and L. L. Larson. 1977. Digestible protein in range forages as an index to nutrient content and animal response. Science Series 29. Fort Collins: Colorado State University Range Science Department.

Cordell, Linda S. 1980. Prehistoric climate and agriculture. In Tijeras Canyon: Analysis of the past, ed. Linda S. Cordell, pp. 60-70. Albuquerque: University of New Mexico Press.

Coues, Elliott, ed. 1893. The history of the Lewis and Clark Expedition, vol. 1. New York: Francis P. Harper.

_____. 1897. The manuscript journals of Alexander Henry and of David Thompson, 1799-1814, vol. 2. New York: Francis P. Harper.

_____. 1898. The journal of Jacob Fowler. New York: Francis P. Harper.

Coupland, Robert T. 1958. The effects of fluctuations in weather upon the grasslands of the Great Plains. Botanical Review 24(5):273-317.

Court, Arnold. 1974. The climate of the conterminous United States. In World survey of climatology, vol. 11, Climates of North America, ed. R. A. Bryson and F. K. Hare, pp. 193-266. Amsterdam: Elsevier Scientific Publishing Company.

Covert, Herbert H., and Richard F. Kay. 1981. Dental micro-wear and diet: Implications for determining the feeding behaviors of extinct primates, with a comment on the dietary pattern of Sivapithecus. American Journal of Physical Anthropology 55:331-36.

Damon, P. E., C. W. Ferguson, A. Long, and E. I. Wallick. 1974. Dendrochronologic calibration of the radiocarbon time scale. American Antiquity 39(2):350-66.

Davies, Owen L., ed. 1961. Statistical methods in research and production. New York: Hafner.

Dean, Jeffrey S., and William J. Robinson. 1977. Dendroclimatic variability in the American Southwest, A.D. 680 to 1970. Final report to the National Park Service, U.S. Department of the Interior (National Technical Information Service PB-266-340).

DeNiro, Michael J., and Samuel Epstein. 1978a. Carbon isotopic evidence for different feeding patterns in two hyrax species occupying the same habitat. Science 201:906-8.

_____. 1978b. Influence of diet on the distribution of carbon isotopes in animals. Geochimica et Cosmochimica Acta 42:495-506.

Dibble, David S., and Dessamae Lorrain. 1968. Bonfire Shelter: A stratified bison kill site, Val Verde County, Texas. Miscellaneous Paper 1. Austin: University of Texas, Texas Memorial Museum.

Dibble, Harold L., and John C. Whittaker. 1981. New experimental evidence on the relation between percussion flaking and flake variation. Journal of Archaeological Science 8:283-96.

Dietz, Albert A. 1946. Composition of normal bone marrow in rabbits. Journal of Biological Chemistry 165(2):505-11.

_____. 1949. Chemical composition of normal bone marrow. Archives of Biochemistry 23(2):211-21.

Dillehay, Tom D. 1974. Late Quaternary bison population changes on the Southern Plains. Plains Anthropologist 19:180-96.

Doliner, L. H., and P. A. Jolliffe. 1979. Ecological evidence concerning the adaptive significance of the C_4 dicarboxylic acid pathway of photosynthesis. Oecologia (Berlin) 38:23-34.

Dore, W. G. 1960. Silica deposits in leaves of Canadian grasses. Annual Meeting Proceedings 6. Guelph, Ont.: Canadian Society of Agronomy.

Dorsey, J. Owen. 1884. Omaha sociology. Bureau of American Ethnology Annual Report 3. Washington, D.C.: Smithsonian Institution.

Downton, W. J. S. 1975. The occurrence of C_4 photosynthesis among plants. Photosynthetica 9(1):96-105.

Draper, H. H. 1977. The aboriginal Eskimo diet in modern perspective. American Anthropologist 79:309-16.

Driesch, Angela von den. 1976. A guide to the measurement of animal bones from archaeological sites. Peabody Museum Bulletin 1. Cambridge: Harvard University, Peabody Museum of Archaeology and Ethnology.

Duffield, Lathel Flay. 1970. Some Panhandle Aspect sites: Their vertebrates and paleoecology. Ph.D. diss., Department of Anthropology, University of Wisconsin.

_____. 1973. Aging and sexing the post-cranial skeleton of bison. Plains Anthropologist 18(60):132-39.

Earl of Southesk. 1969. Saskatchewan and the Rocky Mountains: A diary and narrative of travel, sport, and adventure, during a journey through the Hudson's Bay Company's territories, in 1859 and 1860. Rutland, Vt.: Charles E. Tuttle.

Eickmeier, W. G. 1978. Photosynthetic pathway distributions along an aridity gradient in Big Bend National Park, and implications for enhanced resource partitioning. Photosynthetica 12(3):290-97.

Empel, Wojciech, and Tadeusz Roskosz. 1963. Das Skelett der Gliedmassen des Wisents, Bison bonasus (Linnaeus, 1758). Acta Theriologica 7(13):259-99.

Ewers, John C. 1955. The horse in Blackfoot Indian culture. Bureau of American Ethnology Bulletin 159. Washington, D.C.: Smithsonian Institution.

_____. 1958. The Blackfeet. Norman: University of Oklahoma Press.

Field, R. A. 1976. Increased animal protein production with mechanical deboners. World Review of Animal Production 12(1):61-73.

Field, R. A., H. M. Sanchez, T. H. Ji, Y.-O. Chang, and
F. C. Smith. 1978. Amino acid analysis and acrylamide
gel electrophoresis patterns of bovine hemopoietic marrow. Journal of Agriculture and Food Chemistry
26(4):851-54.

Field, R. A., L. R. Sanchez, J. E. Kunsman, and W. G. Kruggel. 1980. Heme pigment content of bovine hemopoietic
marrow and muscle. Journal of Food Science
45(5):1109-12.

Fletcher, Alice C., and Francis La Flesche. 1972. The Omaha
tribe, vol. 1. Lincoln: University of Nebraska Press.

Food and Agriculture Organization. 1977. Dietary fats and
oils in human nutrition. Food and Nutrition Paper 3.
Rome: United Nations Food and Agriculture Organization.

Francis, R. C., and M. Campion. 1972. Statistical analysis
of intraseasonal herbage dynamics in a variety of grassland communities. US/IBP Grassland Biome Technical
Report 141. Fort Collins: Colorado State University.

Franzmann, Albert W., and Paul D. Arneson. 1976. Marrow fat
in Alaskan moose femurs in relation to mortality factors. Journal of Wildlife Management 40(2):336-39.

Frison, George C. 1970. The Glenrock Buffalo Jump, 48CO304:
Late prehistoric period buffalo procurement and butcher-
ing on the northwestern Plains. Plains Anthropologist
Memoir 7:1-66.

_____. 1973. The Wardell Buffalo Trap 48 SU 301: Communal
procurement in the Upper Green River basin, Wyoming.
Anthropological Paper 48. Ann Arbor: University of
Michigan Museum of Anthropology.

_____, ed. 1974. The Casper Site. New York: Academic
Press.

_____. 1978. Prehistoric hunters of the High Plains. New
York: Academic Press.

Frison, George C., M. Wilson, and D. J. Wilson. 1976. Fossil
bison and artifacts from an early Altithermal period arroyo trap in Wyoming. American Antiquity 41(1):28-57.

Fuller, W. A. 1961. The ecology and management of the
American bison. La Terre et la Vie 108(2-3):286-304.

Gay, Charles W., Jr., and Don D. Dwyer. 1970. New Mexico range plants. Cooperative Extension Service Circular 374. Las Cruces: New Mexico State University.

Gelfand, R. A., R. G. Hendler, and R. S. Sherwin. 1979. Dietary carbohydrate and metabolism of ingested protein. Lancet 1:65-68.

Getty, Robert. 1975. Sisson and Grossman's the anatomy of the domestic animals. 5th ed. Philadelphia: W. B. Saunders.

Goodhart, Robert S., and Maurice E. Shils, eds. 1980. Modern nutrition in health and disease. 6th ed. Philadelphia: Lea and Febiger.

Grinnell, George Bird. 1972. The Cheyenne Indians, vol. 1. Lincoln: University of Nebraska Press.

Gunnerson, Dolores A. 1972. Man and bison on the Plains in the protohistoric period. Plains Anthropologist 17(55):1-10.

Guthrie, Helen A. 1975. Introductory nutrition. 3d ed. Saint Louis: C. V. Mosby.

Halloran, Arthur F. 1961. American bison weights and measurements from the Wichita Mountains Wildlife Refuge. Proceedings of the Oklahoma Academy of Science 41:212-18.

_____. 1968. Bison (Bovidae) productivity on the Wichita Mountains Wildlife Refuge, Oklahoma. Southwestern Naturalist 13(1):23-26.

Halloran, Arthur F., and Bryan P. Glass. 1959. The carnivores and ungulates of the Wichita Mountains Wildlife Refuge, Oklahoma. Journal of Mammalogy 40(3):360-70.

Hammond, George P., and Agapito Rey. 1966. The rediscovery of New Mexico, 1580-1594. Albuquerque: University of New Mexico Press.

Handreck, K. A., and L. H. P. Jones. 1968. Studies of silica in the oat plant. IV, Silica content of plant parts in relation to stage of growth, supply of silica, and transpiration. Plant and Soil 29(3):449-59.

Harris, Dave. 1945. Symptoms of malnutrition in deer. Journal of Wildlife Management 9(4):319-22.

Hatch, M. D., C. R. Slack, and H. S. Johnson. 1967. Further studies on a new pathway of photosynthetic carbon dioxide fixation in sugar-cane and its occurrence in other plant species. Biochemistry Journal 102:417-22.

Healy, W. B. 1973. Nutritional aspects of soil ingestion by grazing animals. In Chemistry and biochemistry of herbage, vol. 1, ed. G. W. Butler and R. W. Bailey, pp. 567-88. New York: Academic Press.

Herbel, C. H., F. N. Ares, and R. A. Wright. 1972. Drought effects on a semidesert grassland range. Ecology 53(6):1084-93.

Houghton, Frank E. 1974. The climate of New Mexico. In Climates of the states, vol. 2, Western states, by Officials of the United States National Oceanic and Atmospheric Administration, pp. 794-810. Port Washington, N.Y.: Water Information Center.

Humphrey, Robert R. 1958. The desert grassland: A history of vegetational change and an analysis of causes. Botanical Review 24(4):193-252.

_____. 1962. Range ecology. New York: Ronald Press.

Hungate, R. E. 1975. The rumen microbial ecosystem. Annual Review of Ecology and Systematics 6:39-66.

Hunsley, R. E., R. L. Vetter, E. A. Kline, and W. Burroughs. 1971. Effects of age and sex on quality, tenderness and collagen content of bovine longissimus muscle. Journal of Animal Science 33(5):933-38.

Irving, L., K. Schmidt-Nielsen, and N. S. B. Abrahamsen. 1957. On the melting points of animal fats in cold climates. Physiological Zoology 30(2):93-105.

Jelinek, Arthur J. 1967. A prehistoric sequence in the Middle Pecos Valley, New Mexico. Anthropological Paper 31. Ann Arbor: University of Michigan Museum of Anthropology.

Jenness, Diamond. 1923. Report of the Canadian Arctic Expedition, 1913-1918 (Southern Party, 1913-1916), vol. 12, The Copper Eskimos. Ottawa: F. A. Acland.

Jochim, Michael A. 1976. Hunter-gatherer subsistence and settlement: A predictive model. New York: Academic Press.

_____. 1981. Strategies for survival. New York: Academic Press.

Johnston, A., L. M. Bezeau, and S. Smoliak. 1967. Variation in silica content of range grasses. Canadian Journal of Plant Science 47(1):65-71.

Jones, L. H. P., and K. A. Handreck. 1967. Silica in soils, plants, and animals. Advances in Agronomy 19:107-49.

Kelley, Vincent C. 1971. Geology of the Pecos country, southeastern New Mexico. Memoir 24. Socorro: New Mexico State Bureau of Mines and Mineral Resources.

Knox, J. H., J. W. Benner, and W. E. Watkins. 1941. Seasonal calcium and phosphorus requirements of range cattle, as shown by blood analyses. Agricultural Experiment Station Bulletin 282. Las Cruces: New Mexico College of Agriculture and Mechanic Arts.

Kobryn, Henryk. 1973. The thorax in European bison and other ruminants. Acta Theriologica 18(17):313-41.

Kobrynczuk, Franciszek. 1976. Joints and ligaments of hindlimbs of the European bison in its postnatal development. Acta Theriologica 21(4):37-100.

Kobrynczuk, Franciszek, and Henryk Kobryn. 1973. Postembryonic growth of bones of the autopodia in the European bison. Acta Theriologica 18(16):289-311.

Küchler, A. W. 1964. Potential natural vegetation of the conterminous United States. Special publication 36. Washington, D.C.: American Geographical Society.

Kunsman, J. E., M. A. Collins, R. A. Field, and G. J. Miller. 1981. Cholesterol content of beef bone marrow and mechanically deboned meat. Journal of Food Science 46:1785-88.

LaMarche, Valmore C., Jr. 1974. Paleoclimatic inferences from long tree-ring records. Science 183:1043-48.

Lamb, H. H. 1977. Climate: Present, past and future, vol. 2, Climatic history and the future. London: Methuen.

Land, L. S., E. L. Lundelius, Jr., and S. Valastro, Jr. 1980. Isotope ecology of deer bones. Palaeogeography, Palaeoclimatology, Palaeoecology 32:143-51.

Lee, Richard B. 1972. The !Kung Bushmen of Botswana. In *Hunters and gatherers today*, ed. M. G. Bicchieri, pp. 327-68. New York: Holt, Rinehart and Winston.

Lewin, Joyce, and B. E. F. Reimann. 1969. Silicon and plant growth. *Annual Review of Plant Physiology* 20:289-304.

Link, B. A., R. W. Bray, R. G. Cassens, and R. G. Kauffman. 1970a. Lipid deposition in bovine skeletal muscle during growth. *Journal of Animal Science* 30:6-9.

_____. 1970b. Fatty acid composition of bovine subcutaneous adipose tissue lipids during growth. *Journal of Animal Science* 30:722-25.

_____. 1970c. Fatty acid composition of bovine skeletal muscle lipids during growth. *Journal of Animal Science* 30:726-31.

Link, B. A., R. G. Cassens, R. G. Kauffman, and R. W. Bray. 1970. Changes in the solubility of bovine muscle proteins during growth. *Journal of Animal Science* 30:10-14.

Long, Austin, and Bruce Rippeteau. 1974. Testing contemporaneity and averaging radiocarbon dates. *American Antiquity* 39:205-15.

Lorrain, Dessamae. 1968. Analysis of the bison bones from Bonfire Shelter. In *Bonfire Shelter: A stratified bison kill site, Val Verde County, Texas*, by David Dibble and Dessamae Lorrain, pp. 77-138. Miscellaneous Paper 1. Austin: University of Texas, Texas Memorial Museum.

Lott, Dale F. 1974. Sexual and aggressive behavior of adult male American bison (*Bison bison*). In *The behaviour of ungulates and its relation to management*, vol. 1, ed. V. Geist and F. Walther, pp. 382-94. Publication 24. Morges, Switzerland: International Union for Conservation of Nature and Natural Resources.

_____. 1979. Dominance relations and breeding rate in mature male American bison. *Zeitschrift für Tierpsychologie* 49:418-32.

Lyman, R. Lee. 1978. Prehistoric butchering techniques in the Lower Granite Reservoir, southeastern Washington. *Tebiwa* 13:1-25.

Lynott, Mark J. 1980. Prehistoric bison populations of northcentral Texas. Texas Archaeological Society Bulletin 50:89-101.

McHugh, Tom. 1958. Social behavior of the American buffalo (Bison bison bison). Zoologica 43(1):1-40.

_____. 1972. The time of the buffalo. Lincoln: University of Nebraska Press.

Marcy, Randolph B. 1863. The prairie traveler: A handbook for overland expeditions. London: Trubner.

Marks, Stuart A. 1976. Large mammals and a brave people: Subsistence hunters in Zambia. Seattle: University of Washington Press.

Martin, John T. 1977. Fat reserves of the wild rabbit, Oryctolagus cuniculus (L.). Australian Journal of Zoology 25:631-39.

Mason, Otis Tufton. 1894. North American bows, arrows and quivers. Smithsonian Institution Annual Report for 1893, pp. 631-80.

Matthews, Washington. 1877. Ethnography and philology of the Hidatsa Indians. Miscellaneous Publication 7. Washington, D.C.: United States Geological and Geographical Survey.

Meagher, Margaret M. 1973. The bison of Yellowstone National Park. Scientific Monograph Series 1. Washington, D.C.: National Park Service.

Medicine Crow, Joe. 1978. Notes on Crow Indian buffalo jump traditions. Plains Anthropologist Memoir 14:249-53.

Meng, M. S., G. C. West, and L. Irving. 1969. Fatty acid composition of caribou bone marrow. Comparative Biochemistry and Physiology 30:187-91.

Merwe, N. J. van der, A. C. Roosevelt, and J. C. Vogel. 1981. Isotopic evidence for prehistoric subsistence change at Parmana, Venezuela. Nature 292:536-38.

Merwe, N. J. van der, and J. C. Vogel. 1978. ^{13}C content of human collagen as a measure of prehistoric diet in woodland North America. Nature 276:815-16.

Minnis, Paul E. 1981. Economic and organizational responses to food stress by non-stratified societies: An example from prehistoric New Mexico. Ph.D. diss., Department of Anthropology, University of Michigan.

Minson, D. J., and M. N. McLeod. 1970. The digestibility of temperate and tropical grasses. Proceedings of the International Grassland Congress 11:719-22.

Misra, Gopa, and K. P. Singh. 1978. Some aspects of physiological ecology of C_3 and C_4 grasses. In Glimpses of ecology, ed. J. S. Singh and B. Gopal, pp. 201-5. Jaipur, India: International Scientific Publications.

Munro, H. N. 1964. General aspects of the regulation of protein metabolism by diet and by hormones. In Mammalian protein metabolism, vol. 1, ed. H. N. Munro and J. B. Allison, pp. 381-481. New York: Academic Press.

National Research Council. 1969. United States-Canadian tables of feed composition. Publication 1684. Washington, D.C.: National Academy of Sciences.

_____. 1971. Atlas of nutritional data on United States and Canadian feeds. Washington, D.C.: National Academy of Sciences.

Nelson, Enoch W. 1934. The influence of precipitation and grazing upon black grama grass range. Technical Bulletin 409. Washington, D.C.: U.S. Department of Agriculture.

Nelson, Richard K. 1973. Hunters of the northern forest: Designs for survival among the Alaskan Kutchin. Chicago: University of Chicago Press.

Neuenschwander, L. F., S. H. Sharrow, and H. A. Wright. 1975. Review of tobosa grass (Hilaria mutica). Southwestern Naturalist 20(2):255-63.

Neumann, A. L., and Roscoe R. Snapp. 1969. Beef cattle. 6th ed. New York: John Wiley.

Newlin, H. E., and C. M. McCay. 1948. Bone marrow for fat storage in rabbits. Archives of Biochemistry 17(1):125-28.

Nichols, Robert G., and Michael R. Pelton. 1972. Variations in fat levels of mandibular cavity tissue in white-tail deer (Odocoileus virginianus) in Tennessee. Southeastern Association of Game and Fish Commissioners, Annual Conference Proceedings 26:57-68.

_____. 1974. Fat in the mandibular cavity as an indicator of condition in deer. Southeastern Association of Game and Fish Commissioners, Annual Conference Proceedings 28:540-48.

Nietschmann, Bernard. 1973. Between land and water: The subsistence ecology of the Miskito Indians, eastern Nicaragua. New York: Seminar Press.

Nordan, H. C., I. McT. Cowan, and A. J. Wood. 1968. Nutritional requirements and growth of black-tailed deer, Odocoileus hemionus columbianus, in captivity. Symposium of the Zoological Society of London 21:89-96.

Novakowski, N. S. 1965. Cemental deposition as an age criterion in bison, and the relation of incisor wear, eye-lens weight, and dressed bison carcass weight to age. Canadian Journal of Zoology 43:173-78.

Olsen, Stanley J. 1960. Postcranial skeletal characters of Bison and Bos. Harvard University Paper 35(4). Cambridge: Peabody Museum of Archaeology and Ethnology.

Opler, Morris Edward. 1965. An Apache life-way. New York: Cooper Square Publishers.

Paulsen, Harold A., Jr., and Fred N. Ares. 1962. Grazing values and management of black grama and tobosa grasslands and associated shrub ranges of the Southwest. Forest Service Technical Bulletin 1270. Washington, D.C.: U.S. Department of Agriculture.

Payne, Sebastian. 1969. A metrical distinction between sheep and goat metacarpals. In The domestication and exploitation of plants and animals, ed. Peter J. Ucko and G. W. Dimbleby, pp. 295-306. Chicago: Aldine.

Peden, Donald G. 1972. The trophic relations of Bison bison to the shortgrass plains. Ph.D. diss., Ecology, Colorado State University.

_____. 1976. Botanical composition of bison diets on shortgrass plains. American Midland Naturalist 96(1):225-29.

Peret, Jean, and Raymond Jacquot. 1972. Nitrogen excretion in complete fasting and on a nitrogen-free diet-- endogenous nitrogen. In Protein and amino acid functions, ed. E. J. Bigwood, pp. 73-118. Oxford: Pergamon Press.

Peters, H. F. 1958. A feedlot study of bison, cattalo and Hereford calves. Canadian Journal of Animal Science 38:87-90.

Peterson, Rolf O. 1977. Wolf ecology and prey relationships on Isle Royale. Scientific Monograph Series 11. Washington, D.C.: National Park Service.

Phillips, Paul C., ed. 1940. Life in the Rocky Mountains, by W. A. Ferris. Denver: Old West Publishing Company.

Phillips Petroleum Company. 1963. Pasture and range plants. Bartlesville, Okla.: Phillips.

Pieper, Rex D. 1977. Effects of herbivores on nutrient cycling and distribution. United States/Australia Rangeland Panel Proceedings 2:249-75.

Pieper, R. D., et al. 1978. Chemical composition and digestibility of important range grass species in south-central New Mexico. Agricultural Experiment Station Bulletin 662. Las Cruces: New Mexico State University.

Point, Nicolas. 1967. Wilderness kingdom. New York: Holt, Rinehart and Winston.

Pond, Caroline M. 1978. Morphological aspects and the ecological and mechanical consequences of fat deposition in wild vertebrates. Annual Review of Ecology and Systematics 9:519-70.

Raab, L. M., R. F. Cande, and D. W. Stahle. 1979. Debitage graphs and archaic settlement patterns in the Arkansas Ozarks. Mid-Continental Journal of Archaeology 4(2):167-82.

Ralph, E. K., H. N. Michael, and M. C. Han. 1973. Radiocarbon dates and reality. MASCA Newsletter 9:1-20.

Ransom, A. Brian. 1965. Kidney and marrow fat as indicators of white-tailed deer condition. Journal of Wildlife Management 29(2):397-98.

Rasmussen, J. L., G. Bertolin, and G. F. Almeyda. 1971. Grassland climatology of the Pawnee Grassland. US/IBP Grassland Biome Technical Report 127. Fort Collins: Colorado State University.

Ratcliffe, Philip R. 1980. Bone marrow fat as an indicator of condition in Roe deer. Acta Theriologica 25(26):333-40.

Read, Catherine E. 1971. Animal bones and human behavior: Approaches to faunal analysis in archeology. Ph.D. diss., Department of Anthropology, University of California, Los Angeles.

Redetzke, Keith A., and Leonard F. Paur. 1975. Long-term grazing intensity data from the Great Plains. US/IBP Grassland Biome Technical Report 272. Fort Collins: Colorado State University.

Reher, Charles A. 1974. Population study of the Casper Site bison. In The Casper Site, ed. George C. Frison, pp. 113-24. New York: Academic Press.

Reher, Charles A., and George C. Frison. 1980. The Vore Site, 48CK302, a stratified buffalo jump in the Wyoming Black Hills. Plains Anthropologist Memoir 16:1-190.

Rice, R. W., J. G. Morris, B. T. Maeda, and R. L. Baldwin. 1974. Simulation of animal functions in models of production systems: Ruminants on the range. Federation Proceedings 33(2):188-95.

Richardson, D. P., A. H. Wayler, N. S. Scrimshaw, and V. R. Young. 1979. Quantitative effect of an isoenergetic exchange of fat for carbohydrate on dietary protein. American Journal of Clinical Nutrition 32:2217-26.

Riney, T. 1955. Evaluating condition of free-ranging red deer (Cervus elaphus), with special reference to New Zealand. New Zealand Journal of Science and Technology 36(5):429-63.

Robbins, Charles T., and Barbara L. Robbins. 1979. Fetal and neonatal growth patterns and maternal reproductive effort in ungulates and subungulates. American Naturalist 114(1):101-16.

Rodgers, Joseph D. 1966. Seasonal protein content of some important range grasses in Lynn County, Texas. M.S. thesis, Range Management, Texas Technological College.

Roe, Frank G. 1972. The North American buffalo. 2d ed. New-
ton Abbot: David and Charles.

Rogers, Edward S. 1972. The Mistassini Cree. In Hunters and
gatherers today, ed. M. G. Bicchieri, pp. 90-137. New
York: Holt, Rinehart and Winston.

Rose, Martin R., Jeffrey S. Dean, and William J. Robinson.
1981. The past climate of Arroyo Hondo, New Mexico,
reconstructed from tree rings. Arroyo Hondo Archaeological Series 4. Santa Fe: School of American
Research.

Sanchez, W. A., and J. E. Kutzbach. 1974. Climate of the
American tropics and subtropics in the 1960s and possible comparisons with climatic variations of the last
millennium. Quaternary Research 4:128-35.

Secoy, F. R. 1953. Changing military patterns on the Great
Plains. American Ethnological Society Monograph 21.
Seattle: University of Washington Press.

Shelford, Victor E. 1963. The ecology of North America. Ur-
bana: University of Illinois Press.

Shipman, Pat. 1981. Life history of a fossil. Cambridge:
Harvard University Press.

Sim, A. J. W., et al. 1979. Glucose promotes whole-body protein synthesis from infused aminoacids in fasting man.
Lancet 1:68-72.

Sims, Phillip L., and J. S. Singh. 1978. The structure and
function of ten western North American grasslands. III,
Net primary production, turnover and efficiencies of energy capture and water use. Journal of Ecology
66:573-97.

Sinclair, A. R. E. 1974a. The natural regulation of buffalo
populations in East Africa. I, Introduction and resource
requirements. East African Wildlife Journal 12:135-54.

_____. 1974b. The natural regulation of buffalo populations
in East Africa. II, Reproduction, recruitment and
growth. East African Wildlife Journal 12:169-83.

_____. 1974c. The natural regulation of buffalo populations
in East Africa. III, Population trends and mortality.
East African Wildlife Journal 12:185-200.

_____. 1974d. The natural regulation of buffalo populations in East Africa. IV, The food supply as a regulating factor, and competition. East African Wildlife Journal 12:291-311.

_____. 1975. The resource limitation of trophic levels in tropical grassland ecosystems. Journal of Animal Ecology 44(2):497-520.

_____. 1977. The African buffalo: A study of resource limitation of populations. Chicago: University of Chicago Press.

Sinclair, A. R. E., and P. Duncan. 1972. Indices of condition in tropical ruminants. East African Wildlife Journal 10:143-49.

Smiley, Francis E. 1979. An analysis of cursorial aspects of the biomechanics of the forelimb in Wyoming Holocene bison. M.A. thesis, Department of Anthropology, University of Wyoming.

Spedding, C. R. W. 1971. Grassland ecology. Oxford: Clarendon Press.

Speth, John D. 1981. The role of platform angle and core size in hard-hammer percussion flaking. Lithic Technology 10(1):16-21.

Speth, John D., and William J. Parry. 1978. Late prehistoric bison procurement in southeastern New Mexico: The 1977 season at the Garnsey Site. Technical Report 8. Ann Arbor: University of Michigan Museum of Anthropology.

_____. 1980. Late prehistoric bison procurement in southeastern New Mexico: The 1978 season at the Garnsey Site (LA-18399). Technical Report 12. Ann Arbor: University of Michigan Museum of Anthropology.

Speth, John D., and Katherine A. Spielmann. n.d. Energy source, protein metabolism and hunter-gatherer subsistence strategies. In preparation.

Spielmann, Katherine A. 1982. Inter-societal food acquisition among egalitarian societies: An ecological study of Plains-Pueblo interaction in the American Southwest. Ph.D. diss., Department of Anthropology, University of Michigan.

Stefansson, Vilhjalmur. 1925. The friendly arctic. New York: Macmillan.

_____. 1935a. Adventures in diet, part I. _Harper's Magazine_ 171:668-75.

_____. 1935b. Adventures in diet, part II. _Harper's Magazine_ 172:46-54.

_____. 1936. Adventures in diet, part III. _Harper's Magazine_ 172:178-89.

_____. 1944. _Arctic manual_. New York: Macmillan.

Stuiver, Minze. 1982. A high-precision calibration of the AD radiocarbon time scale. _Radiocarbon_ 24(1):1-26.

Sullivan, Charles H., and Harold W. Krueger. 1981. Carbon isotope analysis of separate chemical phases in modern and fossil bone. _Nature_ 292:333-35.

Syvertsen, J. P., G. L. Nickell, R. W. Spellenberg, and G. L. Cunningham. 1976. Carbon reduction pathways and standing crop in three Chihuahuan desert plant communities. _Southwestern Naturalist_ 21(3):311-20.

Tamers, M. A., and F. J. Pearson, Jr. 1965. Validity of radiocarbon dates on bone. _Nature_ 208:1053-55.

Tauber, Henrik. 1981. ^{13}C evidence for dietary habits of prehistoric man in Denmark. _Nature_ 292:332-33.

Taylor, Clara M., and Orrea F. Pye. 1966. _Foundations of nutrition_. 6th ed. New York: Macmillan.

Teeri, J. A., and L. G. Stowe. 1976. Climatic patterns and the distribution of C_4 grasses in North America. _Oecologia_ (Berlin) 23:1-12.

Tindale, Norman B. 1972. The Pitjandjara. In _Hunters and gatherers today_, ed. M. G. Bicchieri, pp. 217-68. New York: Holt, Rinehart and Winston.

Todd, Lawrence C., Jr., and Jack L. Hofman. 1978. A study of the bison mandibles from the Horner and Finley sites: Two Paleoindian bison kills in Wyoming. _Wyoming Contributions to Anthropology_ 1:67-104.

Turner, Jack C. 1979. Adaptive strategies of selective fatty acid deposition in the bone marrow of desert bighorn sheep. _Comparative Biochemistry and Physiology_ 62A:599-604.

Valentine, K. A. 1970. Influence of grazing intensity on improvement of deteriorated black grama range. Agricultural Experiment Station Bulletin 553. Las Cruces: New Mexico State University.

Van Dyne, George M. 1973. Analysis of structure, function and utilization of grassland ecosystems, vol. 2. Unpublished progress report submitted to the National Science Foundation by the Natural Resource Ecology Laboratory, Colorado State University, Fort Collins.

_____. 1975. An overview of the ecology of the Great Plains grasslands with special reference to climate and its impact. US/IBP Grassland Biome Technical Report 290. Fort Collins: Colorado State University.

Von Eschen, G. F. 1961. The climate of New Mexico. Business Information Series 37. Albuquerque: University of New Mexico, Bureau of Business Research.

Voorhies, M. R. 1969. Taphonomy and population dynamics of an early Pliocene vertebrate fauna, Knox County, Nebraska. Contributions to Geology, Special Paper 1. Laramie: University of Wyoming.

Walker, A., H. N. Hoeck, and L. Perez. 1978. Microwear of mammalian teeth as an indicator of diet. Science 201:908-10.

Wallace, J. D., J. C. Free, and A. H. Denham. 1972. Seasonal changes in herbage and cattle diets on sandhill grassland. Journal of Range Management 25(2):100-4.

Warren, Robert J. 1979. Physiological indices for the assessment of nutritional status in white-tailed deer. Ph.D. diss., Fisheries and Wildlife Science, Virginia Polytechnic Institute and State University.

Warren, Robert J., and Roy L. Kirkpatrick. 1978. Indices of nutritional status in cottontail rabbits fed controlled diets. Journal of Wildlife Management 42(1):154-58.

Watt, B. K., and A. L. Merrill. 1963. Composition of foods: Raw, processed, prepared. Handbook 8. Washington, D.C.: U.S. Department of Agriculture.

Weitzner, Bella. 1979. Notes on the Hidatsa Indians based on data recorded by the late Gilbert L. Wilson. Anthropological Paper 56(2). New York: American Museum of Natural History.

Weltfish, Gene. 1965. The lost universe: Pawnee life and culture. Lincoln: University of Nebraska Press.

West, George C., and Diane L. Shaw. 1975. Fatty acid composition of Dall sheep bone marrow. Comparative Biochemistry and Physiology 50B:599-601.

Wheat, Joe Ben. 1972. The Olsen-Chubbuck Site: A Paleo-Indian bison kill. Memoir 26. Washington, D.C.: Society for American Archaeology.

White, Theodore E. 1952. Observations on the butchering technique of some aboriginal peoples, I. American Antiquity 17(4):337-38.

Willard, E. Earl, and Joseph L. Schuster. 1973. Chemical composition of six southern Great Plains grasses as related to season and precipitation. Journal of Range Management 26(1):37-38.

Williams, G. J., III, and J. L. Markley. 1973. The photosynthetic pathway type of North American shortgrass prairie species and some ecological implications. Photosynthetica 7(3):262-70.

Wilmsen, Edwin N. 1970. Lithic analysis and cultural inference: A Paleo-Indian case. Anthropological Paper 16. Tucson: University of Arizona.

Wilson, Gilbert L. 1924. The horse and the dog in Hidatsa culture. Anthropological Paper 15(2). New York: American Museum of Natural History.

Wilson, Michael. 1980. Population dynamics of the Garnsey Site bison. In Late prehistoric bison procurement in southeastern New Mexico: The 1978 season at the Garnsey Site (LA-18399), by John D. Speth and William J. Parry, pp. 88-129. Technical Report 12. Ann Arbor: University of Michigan Museum of Anthropology.

Wing, Elizabeth S., and Antoinette B. Brown. 1979. Paleonutrition. New York: Academic Press.

Wissler, Clark. 1910. Material culture of the Blackfoot Indians. Anthropological Paper 5(1). New York: American Museum of Natural History.

Index